MW00713791

Desktop Computer Animation

Desktop Computer Animation

A Guide to Low-Cost Computer Animation

Gregory MacNicol

Focal Press
Boston London

Focal Press is an imprint of Butterworth–Heinemann.

Recognizing the importance of preserving what has been written, it is the policy of Butterworth–Heinemann to have the books it publishes printed on acid-free paper, and we exert our best efforts to that end.

Library of Congress Cataloging-in-Publication Data

MacNicol, Gregory.
Desktop computer animation : a guide to low-cost computer animation / by Gregory MacNicol.
p. cm.
Includes index.
ISBN 0-240-80065-6
1. Computer animation. I. Title.
TR897.5.M32 1992
006.6—dc20 91-45650
CIP

British Library Cataloguing in Publication Data

MacNicol, Gregory
Desktop Computer Animation : A Guide to
Low-cost Computer Animation
I. Title
006.6
ISBN 0-240-80065-6

Butterworth–Heinemann
80 Montvale Avenue
Stoneham, MA 02180

10 9 8 7 6 5 4 3 2 1

Printed in the United States of America

Contents

CHAPTER 11

CHAPTER 12

CHAPTER 13

CHAPTER 14

Preface

This book is intended as a guide for the computer animator, the neophyte as well as those experienced with the animation process. Now that professional grade software and hardware are accessible and affordable for desktop systems, anyone with the requisite vision and fortitude can turn dreams into viewable realities. What this means to a less technically advantaged animator, artist, or filmmaker is that professional-level storytelling technology is now available. In fact, the advent of the personal computer, coupled with capable animation software, has signaled a new era in the world of animation—it's storytelling technology for "the rest of us."

In the past few years, the range and power of the best and most expensive mainframe computers have been challenged by newer desktop computers, which are smaller and less expensive. These more affordable and thus more available computers now have the capability to produce excellent animation. Indeed, the functional capacity of dedicated graphics workstations has been reduced to plug-in circuit boards so that the personal computer can now be transformed into a graphics workstation with the addition of a single component. In a similar manner, sophisticated and proven software is constantly being ported down to these smaller computational engines from the mainframe world, and the hardware continues to evolve even as it becomes less expensive.

What is a "desktop computer" or "PC"? Here and throughout this book, the hardware platform—that is the actual computer you see or wish to see in your workroom—referred to includes most of the popular systems widely available in today's marketplace. These include the Apple Macintosh series, the IBM series of personal computers (MS-DOS-based '386 systems and up), and Commodore's Amiga series. You should know, however, that the information in this book is not necessarily computer-specific. If, for example, you happen to use a Macintosh, you will benefit as much from this guide as you would if you owned an IBM-AT. The processes of model creation, motion scripting, and exporting images to film or videotape, apply to most available personal computers that have the ability to display high-quality graphic images.

Theoretically, all projects that are suited to computer animation in general can be managed by PC-based computer animation equipment and software. The primary difference is that equipment costs are lower and, for computationally intensive work (such as 3-D rendering of realistic images), production times are invariably slower. Thus, compared to production using mainframe-based systems, an animated sequence will take longer to complete, given an animation project of equal complexity.

Those who come to computer animation typically arrive via a background in either computers or the graphic arts. Because this book attempts to address the perspectives and the obstacles intrinsic to both points of view, those with computer expertise may wish to skip the computer-related technical descriptions. Likewise, those with art production experience may opt to skip particular sections in this book, such as the chapter on film recorders.

Luckily, you needn't be a technical wizard to assemble your own "dream machine." Yet the reader of this book is assumed to be somewhat technically literate—that is, familiar with the basics about how a computer operates, how to run programs, and how to install hardware peripherals. This guide is intended to take you a considerable way down the road toward professional-level animation. Armed with the capability inherent in much of the currently available software and with the information provided in this guide about the expandability of your "personal" computer system, you should be able to custom design an animation system suited to your purposes and desired end product, be it video or film. Indeed, for the ambitious, this guide should provide you all the informational tools necessary to create professional broadcast-quality video.

Once you've developed some confidence using the tools of the computer animator's trade, you'll discover that most animation projects are not extremely complex. The so-called 80/20 rule applies to many endeavors and is also descriptive of the animation process: Most animation tasks are easily accomplished (given the required tools) and can be assigned to the 80% category—that is, workaday production such as simple animated logos and titles. The remaining 20% covers the more creative or technically challenging work. Thus, if most animation can be completed effectively using a PC-based system, those with limited resources should consider leaving the specialized stuff to those with the technical resources to get the job done expeditiously, if not economically. After all, the 80% represents a broad canvas—and will provide the nascent animator with a powerful mix of video, 3-D synthesis, 2-D imagery, and titles.

In fact, the best computer animation is usually the result of creative personnel developing a superior script and capable artwork, and then following up with imaginative execution. Beyond a certain level, the impact of equipment is secondary to talent and craft. This is probably an underlying reason why computer animation on PCs is becoming so popular.

Given the hardware, and reasonably sophisticated software, all you need is an abiding interest in animation and a little technical self-confidence. You need to be able to acquaint yourself thoroughly with the workings of one of the better software animation programs. You will discover that successful low-cost computer animation production requires the tenacity to follow a production through to the end. It's a slow-moving yet rewarding process as you see increments of your ideas drop into place and come to life. And finally, a touch of the artist definitely won't hurt you. After all, animation is basically a visual craft, an artistic endeavor that has never suffered from strong mixtures of talent and imagination.

Topics have been arranged in such a way that the guide can be used as an animator's reference handbook. In fact you will probably find that there is simply too much detailed information to be assimilated in one sitting. That's why we suggest that you read through the whole book first and then focus on the material that will be most useful to you. For example, take the topic of video. Computer animation with video as a production objective requires video expertise (which is provided in Chapters 13 and 14). If you are already knowledgable about the fundamentals of video, you might elect to proceed directly to Chapter 14, on video-to-computer interfacing.

Perhaps instead you are interested in putting together a desktop system from scratch. In this case, you would be well advised to carefully examine the beginning chapters first and then follow up your investigation with a reading of Chapters 7–10, which detail software considerations. If you read the book sequentially, you will encounter occasional redundancies due to the importance of certain topics: Overlapping inevitably occurs in a technology as complex as this. However, enough information is included about each topic so that each chapter can stand on its own without unnecessary diversions to other chapters.

One of the challenges to those involved in desktop computer animation is that virtually every aspect regarding the technology is changing daily. New advances eclipse old techniques, while at the same time long-forgotten methods from the "olden days" are revived in new software updates. A good example of this is the relationship of 2-D and 3-D animation. Most people think high-quality computer animation is primarily a 3-D technology, not realizing that many techniques are essentially 2-D or a mixture of both. As a result, this book focuses primarily on the more technical aspects of 3-D computer animation technology, without trying to exclude the importance of 2-D. Chapter 7 explores 2-D capabilities and, when appropriate, shows how they can be used to augment the 3-D animation process.

Computer animation is a new and still rapidly evolving art form. Practically every aspect of the technology and the media, including its aesthetics and attitude, are changing beyond any one person's ability to formalize it. If you choose, you can view this as an advantage: Because of the burgeoning technological opportunities that are the harvest of so much raw capability, desktop computer animation opens the doorway to a massive and virtually uncharted visual world. The tools of this powerful medium are now in your hands. Indeed, after a careful reading of this book, you should have all the conceptual tools and information necessary to embark on your own computer animation project.

Acknowledgments

I am particularly grateful to the following companies, who provided great assistance in the preparation of this book:

AT&T GSL, Autodesk, Byte-by-Byte, Crystal Graphics, Diamond Computer Systems, Diaquest, JVC, Lasergraphics, Macromind/Paracomp, Micronics, Micropolis, Mitsubishi, NewTek, Panasonic, Sigma, Sony, Storage Dimensions, Truevision, Wacom, Weitek, and Wyse.

Desktop Computer Animation

What Is Computer Animation?

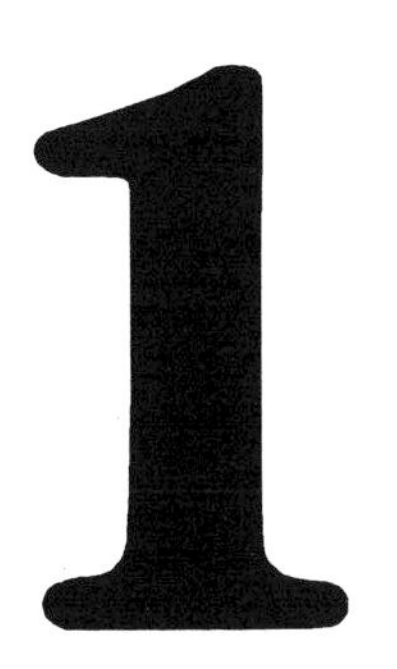

Let's start with a simple truth—animation is illusion. Yes, all the films you see on TV or in the movie theater (animated or otherwise) are nothing more than believable (or not so believable) simulations of movement. Motion picture films are composed of a series of static images, singular time slices capturing unique light configurations that have been frozen in time. When they are arranged in series and played back on a projection screen at the rate of 24 frames per second, you get a two-dimensional translation of three-dimensional live movement.

Normally, motion pictures are projected onto a screen at a rate of 24 images per second—yet, actually, each frame is shown twice and separated by blackness to match our brain's tendency to retain the last perceived image. Video is based on the same principle. A video production is shown 30 times per second—two low-resolution images are projected every 1/60th of a second. Thus, each frame of video or film is simply a fraction of the complete movement. Like a photograph, each image is still—that is, inanimate. Therefore even in the best motion pictures there is an inherent jumpiness, which is an inevitable result of the compilation of the separate, individual images.

After all, motion picture images are simply samplings of real movement that have been reconstituted in some fashion after the fact. As in all histories, important elements are left out in the retelling. In other words, with respect to motion pictures, less is there than meets the eye.

Fortunately, that's not the way people see it. The persistence, or visual memory, of the image acts as a fill-in for a similar succeeding image and thus serves to bridge a sequence of static images. The human eye retains the last perceived image for a brief moment. That's why, as soon as people get caught up in the content, they readily get used to and accept even the crudest of simulations. This is how films can convey the illusion of smooth, uninterrupted movement. Another way of looking at this phenomenon is to say that the human eye is slow to perceive rapid movements—if you don't believe this, ask a magician. A good example is the flip-book animations made by children on pads of paper. Even these simple productions can invoke a believable sense of motion. As the poet Coleridge once put it, the task of the storyteller—regardless of the medium—is to make us "willingly suspend our disbelief."

To create a flip-book "motion picture" from scratch, kids (of all ages) assemble a sequence of static images on small paper pads and arrange them for playback using the thumb as a kind of primitive projector. Regardless of the level of sophistication, all animation is based on this principle.

It's a tedious business when you think about it. At 24 frames per second for film (30 for video), 86,400 pictures are required to produce an hour of film! That's where computers come into the picture. The computer's great strength is in the management of redundancy; that is why it's so well suited to the repetitious work of animation. That's the good news. The not-so-good news is that although the computer provides powerful arithmetical capabilities and accessible mass storage, which facilitate sequential tasks, animation still requires considerable human interaction. Generally, the more complex the animation, the more interaction is required to produce believable animation.

The difference between animation and still imagery is more than moving pictures. As a communication tool, animation is an especially powerful way to present information. An animated bar chart, for example, can convey considerably more information (in a remarkably more compelling fashion) than a static graph can. Via animation, the presentation of the chart is not limited to a simple overview of information but rather dramatizes interrelationships among key components and thus describes how one variable impacts another. The viewer's perceptual and conceptual process can actually be guided as key points are illustrated. In other words, a conventional bar chart can be transformed into a kind of story, one that is not only interesting and informative, but also one that has your ending built in.

Some animators conceive of the animation process as analogous to the movements of actors on a stage set. Indeed, some animation software packages, utilizing the same metaphor, refer to "actors" and "scripts" in their animation methodology. The analogy is apt because it conveys the notion that you, as the animator, are the ultimate producer/director. You set the stage, cast the actors, write and edit the script, manage the lights, and laboriously coordinate the process through to a finished product.

It's a godlike enterprise. Unlike motion picture production, you're unlimited in terms of image creation: You don't have the hassle of personal politics when assigning jobs; nor do you have to manage a large staff to get your tale told. You decide how the story is to be shaped from beginning to end. This autonomy is powerfully attractive. You can steadfastly translate your vision into a finished product that communicates your own unique concepts and vision.

Types of Computer Animation

What is computer animation? What kinds of computer animation are discussed here? The term *computer animation* includes any production of an animation sequence in which a computer is used even casually in the animation process. Computers are used to assist the animator in two distinctly different ways. The computer can be used as a support tool outside the animation process, or it can be used in the actual creation of the individual frames.

In the most casual and least intensive application, the computer is used simply in support of a traditional animation process. It is not used to assist in the production of the animation images, but rather it is used in its role as a manager to handle peripheral details such as script integration, sound and music cues, and the like. After all, the computer is an excellent management tool. Another of its great virtues, from the point of view of the animator, is its ability to store voluminous amounts of data, a practical application when you think about it. At the rate of 86,400 frames per hour, animation data accumulate in a hurry. The computer is also useful in more tangential ways. Computers are typically used by animators for ongoing project management—to log film data, to create and maintain edit lists, and to monitor the status of the various animation sequences.

However, it is in its role as animation assistant that the computer's real power becomes evident. Indeed, the computer is already propelling the animation industry in the direction of new forms and innovative video-production tactics and strategies.

In some ways (always with certain qualifications), the computer is already a potent animator. With guidance (input) from the animator, the computer can automatically create the individual frames that comprise an animation sequence. First, the computer must assimilate key pieces of information about an image—information such as light position, camera-lens focal length, and other constantly recurring details whose repetitious iteration taxes the patience of humans. Second, on the basis of these data, the computer can be used to extrapolate additional images and actually produce the components of an animation sequence. In this regard, the computer's projection and calculation abilities are far beyond the capabilities of most animators: Given the right data, the computer can automatically calculate, for instance, how to make water move and how to make objects fall realistically.

The computer is now a key player in almost every aspect of computer animation. In conventional animation production, where individual plastic *cels* are used, for example, the computer is used to automate the camera's point of view, coordinating movements such as zooms, pans, and rotations. It is also used as a special effects tool to create glows and other starlike effects. Likewise, in clay and miniature model animation, producers rely on the computer to help coordinate accurate, dependable, and predictable camera movements. These kinds of productions can be complicated, especially when (as often happens) live action is added to the animation. In such cases, the computer helps to ensure accuracy of the replicated live-action camera movement, and such accuracy facilitates seamless joining of animation to video or film.

The computer can also be used to synthesize imagery, thereby creating a series of pictures that are eventually incorporated into the animation. Depending on the animator's requirements, either 2-D or 3-D techniques can be used (although they are sometimes used together). 2-D animation uses *objects*, or cels that have extension only in two directions: height and width. In order to add a sense of depth, the animator must extend the drawings to the third dimension and impose shadowing. However, whether for budgetary reasons or as a matter of artistic style, working in only two dimensions can be eminently satisfactory.

For more complex work, the computer can be used to create the third dimension. In this case the animator must first fashion 3-D objects in 3-D space. Creating complex and detailed objects is an arduous process with a remarkable benefit: Camera movement and shadows can be handled automatically by the computer. However, whether the production is two dimensional or embraces the third dimension, the computer is typically used to organize the animation into a sequence of images and to facilitate the process of putting the images onto film or video.

Software is now available that enables the computer to produce photo-realistic images. When the software is merged with hardware that displays the images on a computer monitor (either in TV resolution or better), the picture begins to emerge. Computer graphics software can now create scenes of such extraordinary quality that it is no longer possible to tell that the images are computer-generated animations—they are that realistic and visually believable. This level of realism is no longer the exception—believable hyper-realistic animation is now commonplace. What this means to today's animator is that if something can be imagined it can be rendered as if it were a real object. Thanks to the computer, people can repeatedly live in a world of magical capability.

Animators would all like to see the tedium and attendant drudgery of the animation process obviated by fast, friendly software programs. In the best of all possible worlds, the animator would use the computer as a kind of super tool, an extension of the creative self that responds to simple directions and, via a simple query process, produces complete animation sequences. The animator would type in an animation script and the computer would create all the frames, following the descriptions. As you might imagine, such a script would have to be very explicit in order to produce fully detailed and complete 3-D models. So how does this compare with the real world?

To some extent, these tools are already here, albeit in relatively crude and usually problematic guises. As things now stand, considerable distance divides the animator's ideals from the present capabilities of the computer. We are still a long way from complete automation, but, as you will see, animation systems are starting to approximate some exceedingly intelligent attributes.

Computer animators have differing opinions about how such a perfect system should function. Stridently divergent conceptual platforms and attitudes complicate the development of computer hardware and software: There is no consensus regarding the management of attendant processes such as video, editing, and production. It seems to be axiomatic that greater power requires more difficult and more complex decision making. Issues continually emerge regarding quality, approaches to motion scripting, and the level of involvement of the animator. These issues are reflected in widely varying approaches to the technology. This is the main reason for the variety of approaches taken by the producers of the software available for PCs.

Software Strategies

This book concentrates not on computer animation processes that are merely computer assisted (in which the computer serves as a helper) but on processes in which the computer plays an integral role in the development of animation images—parts of an animation sequence that are in one way or another created or augmented via the computer. This section introduces some of the software strategies that typify what's currently available.

A simple kind of computer-animation software is based on the computer's excellent ability to save and recall images. With this software, the animation process involves making computer-assisted images one by one and saving the individual images so that they can be sequenced later. Usually, in this approach, a program is used to paint images, to scale and rotate them, to cut and paste them where necessary, to incorporate photographs, and finally to create composite images combining "real" and drawn images. These processed images are saved onto a hard disk or other storage medium and later transferred directly onto videotape or film. Systems that operate in this way are called *2-D animation systems* because the artist provides the semblance of a 3-D world. 2-D animators hand draw the appearance of depth, including the telltale cues from shadows. *3-D animators,* in contrast, rely on the computer to create the sense of depth, perspective, and shadows. In 2-D animation mode, the artist (rather than the computer) does the bulk of the imaging.

The computer can assist in the creation of individual images in several basic ways. It is typically used to do the following:

- *Create black-and-white line drawings in three dimensions on paper.* These drawings are subsequently filled in by an animator. Hand-drawn characters may be added later.
- *Add color.* The computer can color-process a black-and-white image and, under operator control, color selected parts.
- *Create script motion.* The computer can generate in-between images averaging the motion between scenes.
- *Manage virtual cameras, lights, shadows, and textures to create an image.* Making the image by computer is called *rendering.*
- *Integrate pictures and computer-generated images, such as words and live video, into the contiguous sequences of a finished animation.*
- *Add special effects to an already-produced animation (on video or film).*

All of these images are intended to be of a higher resolution than is available on the usual text-oriented computer screen. They are quality images intended to be suitable for use in film or video. This book does not focus on systems with which an animation is made to run on a low-resolution computer display intended for text. Rather, it focuses on programs intended for such specific applications as the creation of product demos.

Real-Time versus Single-Frame Animation

It is necessary at the outset to discern the difference between real-time and single-frame capability. Some systems have the ability not only to create an animation sequence and save the whole animation in memory (often RAM), but also they can play the sequence back in *real time*. In other words, the system can run the animation at the rate at which it will ultimately be viewed. Some of these systems provide software control so that you can speed up or slow down an animation sequence to determine the best viewing rate. Of course, such systems have their limitations. For example, due to hardware limitations, sections of the animation may slow down when the detailed images become too complex. Systems capable of real-time animation typically enable direct videotaping from the computer. They thus do not require the complex complement of hardware support equipment typically demanded by single-frame video systems.

In contrast, *single-frame animation* systems require more time to create an image than the 30-frames-per-second rate of video. This would require the computer to calculate and display a new frame uninterrupted each 30 milliseconds—which is no trivial task. With systems set up for single-frame animation, each frame of the animation must be placed onto videotape as it is created. The trade-off for this penalty in terms of added equipment costs and boosted time requirements is vastly superior output quality. Photographic-like quality is possible, and convincing special effects can be fashioned—effects that mimic the output of even the most sophisticated high-end computer animation facilities. Systems such as these allow full creativity on the part of the animator because they generally provide a complete range of colors, superior resolution, and many extra capabilities.

Applications

Professional uses for PC-based animation systems include a wide range of commercial kinds of video production. For instance, much of the work seen on TV is or could be generated on a PC. A whole range of industrial-type video production opportunities are available to the enterprising animator. Even broadcast-quality video production facilities use PCs. For example, these studios often find they need sophisticated titling, such as rotating letters with shadows, that is superior to the simple output of a character generator.

It is worth noting that all three of the big U.S. networks (as well as others) are using desktop systems to create their new season's logos and graphics (see Figure 1–1). Designers create the 3-D images and 2-D backgrounds on low-cost desktop systems and send them via modem to animation production facilities across the country for final treatment

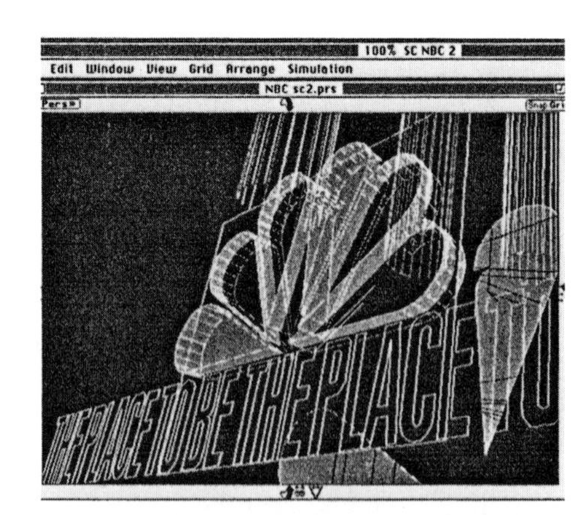

Figure 1–1. *Low-cost computer animation is being used to prototype broadcast network IDs. (Reproduced with permission of Dale Herigstad Design.)*

and translation to video. Today, video-graphics artists who are not computer animation literate have limited functionality in professional video production environments.

Desktop computer animation can be effectively used to create the titles and lettering that typically appears at the beginning (or end) of a story or program. In fact, as many commercial production facilities have already discovered, an inexpensive computer animation system in a broadcast environment is a powerful and low-cost asset in their production arsenal. These systems are used all the time to digitize a frame of video, to make it smaller, to modify it by adding detail, or otherwise to adapt frames for use in some special application. One way to add special effects is to run already-existing video through the animation system to enhance the video image and make it appear as though it were drawn with charcoal or watercolors.

In the industrial market, PC-based computer animation is used frequently (see Figure 1–2), despite the fact that industrial productions usually have limited audience appeal. There are, however, opportunities to aim at a larger audience. Corporations use computer animation productions at trade shows to introduce new products, to provide animated instructions for installations and training, and to create in-house sales demonstrations of new products. If the communication is deemed to be of importance, then the videos must be interesting.

Even a seemingly dull graph can be brought to life with a video-captured background, by a rotation of the graph in a 3-D space, or, more simply, by the addition of shadows. Of course, the graph can be made to move to reflect, in some usual or unusual way, the impact of new data entered to it. Another example of a new use for the industrial video is the 10- to 20-minute video sent to a specialized and very specific kind of customer (such as an engineer) who can learn about a particular product (such as an oscilloscope). The targeted audience can usually learn about a new or unfamiliar product much more rapidly and more enjoyably via a well conceived video than by reading about it. Of course, there's the hidden selling value of emotional approval, which is predictably pro-

Figure 1–2. *Corporate videos are becoming more popular to convey the professional image of the company.*

voked when a high-quality computer as assisted video is the medium for the message.

Applications of animation appear to be limited only by the imagination and the budget. As PC-based computer animation becomes more affordable, new markets are emerging. Law firms, for example, are now using computer animation in court to simulate an interpretation of an accident or technical issue. They are also using animation to demonstrate technical concepts to a jury, who often don't have the time or ability to assimilate written technical documentation.

Another example of an emerging application is in the area of scientific visualization. When complex technical data must be presented to others (to peers or to those less well versed, technically) animation can be used in a convincing and lucid fashion to communicate efficiently new concepts or principles (see Color Plate 1). Government scientists at international conferences, for example, have found it advantageous to use low-cost Amigas to demonstrate the movement of earthquake fault lines in real time. Using a mainframe computer to prepare this kind of simulation would probably take more time because of the great amount of data required and because government systems are typically time-shared. Getting the animation onto video would not be easy, either. In such a case, a low-cost desktop computer is able to play directly into a home VCR the animated seismic activity as it was displayed on the computer screen.

Demand for the visualization of new and/or complex ideas is surging. For example, in the field of architecture, animation is slowly being accepted as a method of presentation. With low-cost animation tools and techniques, CAD data (in a variety of formats) derived from architectural databases can be converted and then animated to produce video presentations. These so-called *fly-throughs* can turn flat schematic line drawings into believable facsimiles of the new building (see Figure 1–3). The benefits are surprising: Structural problems are re-

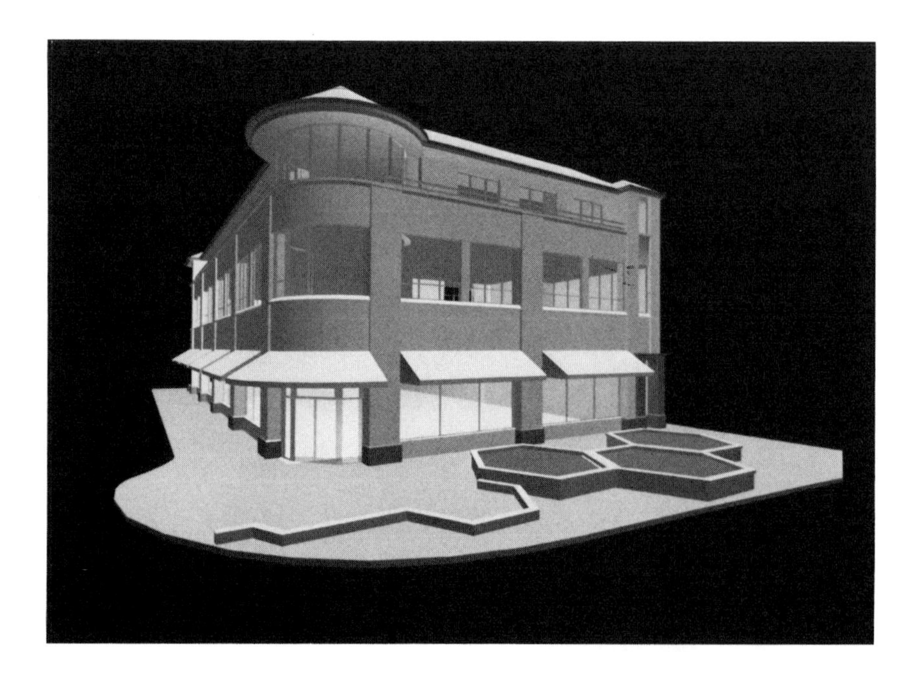

Figure 1–3. *Architectural simulation when animated reveals details otherwise overlooked.*

vealed, incompatibilities in building color are noticed and changed, and aesthetic decisions can be evaluated in the context of the structure. In fact, due to the computer's ability to simulate the sunlight of a particular day at a particular time and latitude, shadows on adjacent buildings can be immediately visualized.

Examples such as this demonstrate the trend toward (and dramatize the need for) enhanced visual communications. Not surprisingly, planners are more effective when they can not only view but also in a real way experience a new project. Many other fields are being impacted similarly: New applications and markets that transcend old-fashioned methods are constantly being created. Because animation technology is still in its infancy, you can be involved in the shaping of new visual frontiers.

It is to this end this book focuses on computer-generated animation techniques. This book explores and explicates the more intensive applications of computer animation—especially those activities in which the computer is instrumental in the creation and control of the individual frames that compose an animation. Topics to be investigated include how the computer is used to assist the animator in the creation of the in-between frames that bridge the movement explicated in *key frames*. Other capabilities will be looked at too, such as computer-assisted coloring, in which an object is painted and shaded so that it has the visual properties of a typical classical animation.

2

The Animation Process: An Overview

Creating even the simplest animations requires careful preliminary homework. You need to know your hardware and software, and you must have at least a rudimentary understanding of animation concepts and terminology. The more you know about the process, the equipment, and your skills beforehand, the better your planning will be and the greater the likelihood of a satisfactory finished product. Many an animated catastrophe has occurred because the last piece would not fit. When all your hard work on an animation project is finished, for example, you don't want to find out that you need a few more seconds somewhere, or a better video signal, or that your ending doesn't work the way you expected.

To achieve competent results in the complex world of animation, you need to be comfortable with whatever hardware and software you will be using. To develop this kind of operational ease, a good preliminary exercise is to explore the range of a system's software functions, while at the same time testing the limits of the hardware. Experienced animators create their own "just-for-the-fun-of-it" animation in order to test a system's capabilities as well as their own ability to operate the software. This is a good way to discover the strengths and limits of your computer system as well as how effectively you interact with it. If possible, it's a good idea to test new software at the store where it is sold—before you purchase it. This gives you a realistic idea of what it takes to create an animation using a particular system.

However, before you consider detailed hardware and software prerequisites, you should gain a sound understanding of the overall animation process. You'll need this basic computer animation background in order to select the appropriate tools for the kind of animation you have in mind. In effect, you've got to bootstrap your way into the process: You've got to decide what hardware is most appropriate, and in order to do that, you need to know what software you'll be running. You need to know what questions to ask so that you can evaluate and select software that's appropriate to your needs.

Just as with most large tasks, animations are best produced by dividing the job into a sequence of subtasks or steps. Before the work begins, *preproduction* determines that all the pieces of the intended animation will fit predictably and neatly in an efficient path. Because the final product is to be a visual presentation, it makes sense that the outline of your proposed animation should also be a visual representation. That's why we recommend that all animation projects begin with what's called a *storyboard*—a series of graphic images, similar to a succession of simple cartoons, that depicts the major events of the work (storyboarding is covered in more detail in Chapter 3). A good storyboard also conveys a sense of the timing as well as the key visual components of your animation. With this graphic outline, each aspect of the animation process can be broken down into its individual segments and production requirements for each can be more easily visualized and clarified.

After the preplanning and storyboarding is complete, production can begin. There are four basic parts or phases to the production of an animation:

- modeling
- rendering
- adding motion
- transferring onto video or film

Production Overview

All animated sequences are composed of *objects*. In a simple animation, for example, we might include three elements: a boy, a bat, and a ball. All three are, in the lexicon of the animator, objects. Of course, before you breathe life into the cast of characters, they must be created: This process is called *modeling*. In order to create a 3-D model, the animator must combine cylinders and other geometric shapes and then contort and *glue* them into a realistic *wire-frame* representation of the envisioned object.

Raw 3-D objects (digital models) need finish and polish to imbue them with the subtle visual cues that connote realism to the eye. During the *rendering* process, 3-D objects or models are shaded with lights designated by the animator—each scene in the animation is assigned a camera angle. Also at this stage, each object's surface features are defined with appropriate colors and surface characteristics.

Therefore, after your objects have been satisfactorily modeled, the next step is to render them. The model is *read* by the rendering part of the program and dissected for the best and most accurate way of displaying the model realistically on the screen. This process depends on many details, and completing a rendered image can take from 1 minute to several hours.

Assuming all has gone well in the rendering, an object can then be instructed to move: This process is called *motion scripting*. Many animation software products allow you to move the camera (adjust the viewpoint of the camera) as well as the angle and the color of the lights. Most programs also allow you to make adjustments to the scale and the positioning of objects. Better animation software products also allow you to control the speed of the action (acceleration) and can automatically calculate in-between frames based on the beginning and end of a given animated sequence.

Finally, after your animation has been planned (storyboarded), modeled, rendered, and motion scripted, it can be *transferred* to video or film, depending on the purposes of your production and the audience the work is intended to serve.

3-D Modeling. A 3-D object possesses apparent height, width, and depth. So if a 2-D object, such as a lettered logo, is to be enhanced with 3-D characteristics, it must be *extruded;* that is, it must be given depth. Most modelers are capable of providing a variety of 2-D fonts that can be used as a starting point. After selection of a font and size, the text can be typed into the work space (see Figure 2–1a). At this point the dimension of depth can be incorporated into the lettering by extruding the logo into a true 3-D object (see Figure 2–1b).

This is where even a relatively simple animation such as a "flying logo" leaves the mundane world of terra firma. Advanced modelers usually provide complex functions beyond simple extrusion. A 3-D object can be created and then *cut,* or subtracted from, by another object. Imagine, for example, that you've constructed a wheel and that you want to add a roughly shaped axle to its midpoint. After the wheel has been fashioned, a squarelike object can be placed in the wheel's center and then subtracted from the wheel. Another way to make the same object is to form a circle, then a square, and place the square in the center of the circle. After the square is subtracted from the circle, the wheel can be extruded to imbue the wheel-like object with a sense of depth.

More sophisticated modelers usually offer more than one way to create an object. A 3-D object can be formed out of polygons (lines that form the 2-D object) or it can be shaped from *splines,* or curved lines that are perfectly rounded. Thus a circle made of polygons would have small telltale facets or lines that when joined together are still perceptible when viewed close up. In contrast, a spline-made circle is perfectly round and will appear so even upon close-up examination (see Figure 2–2). Each method has its advantages and drawbacks. Although splines require more calculations and, hence, are slower to render, spline-fashioned objects are preferable in some situations. For example, constructing a square block from splines instead of polygons allows the animator to bend, twist, and warp the block into any conceivable eccentric shape and to animate the distorted object too. Newer animation programs offer both methods of model creation.

After an object (or set of objects) has been modeled, it can be graced with color and surface characteristics. For example, the logo we have already blocked out can be colored and made to appear highly reflective (see Figure 2–1c). Color limits depend on the capabilities of a given system's hardware. The best systems allow for up to 24 bits of color depth (which translates into 16.7 million useable colors). While that

a

b

c

d

e

Figure 2–1. *Creating a typical flying logo consists of several steps: (a) First, the 2-D outlines of the shape are created. (b) Then the outlines are extruded into 3-D shapes. (c) Surface characteristics are assigned to the 3-D objects. (d) Background objects or images are incorporated, lights are positioned, and a camera position is assigned. (e) Finally, the image can be rendered with automatic shadow generation at a specific resolution.*

may seem like an extravagant array, an animator working with objects that are to be fully shaded and rendered needs a subtle range of gradations for each color. This is especially important during animation, when subtle color differences tend to scintillate if there are not enough available colors.

Surface characteristics, which determine how an object responds to and reflects light, can be assigned a graded value that ranges from diffuse to specular to flat. If, for instance, surface characteristics are set for complete diffusion, an object will appear dull and plasticlike. If the specularity value is high, it will have a shiny, reflective appearance like that of shiny metal. A company logo, for example, should be easily readable and so should be assigned fairly flat surface characteristics.

Most of the better 3-D animation programs integrate modeling and rendering features. You should be aware, however, that some low-cost programs may provide only one of these functions. Because no standard model format has yet evolved, this can be a problem when transferring work to a different system using a different software product.

Figure 2–2. *Curved lines (such as in the letter G, for example) appear smoother when made with splines (right) than with sequential (left).*

Rendering. Rendering brings a 3-D model to life. Light sources are added and positioned; the camera's angle or viewpoint is determined; and special details are added to embellish the model with photo-realistic detail (see Figure 2–1d). Visual screens or *maps* can be superimposed to add a wealth of visual minutiae. For example, let's say you created a wall and want to make it look like the inside of a prehistoric cave. You can take a photograph of cave art, run it through a scanner or capture it with video, and save it as a digital image. This image can then be used as a map that, when projected onto the wall, will appear as if it were painted on the wall—complete with any distortion resulting from the wall's curvature or surface irregularities. If you want to avoid distortion from whatever is serving as a screen, the image can be *wrapped* (like a present) instead. Other types of mapping, such as *bump mapping,* can add remarkably realistic attributes to a completed image. A good example of bump mapping is the addition of little textured orange-peel–like irregularities to an otherwise smooth sphere.

Perhaps the most compelling rendered attributes are shadows. Because shadows have controllable qualities, images can be rendered that have distinctive impact and originality. The sharpness of the edges, the granularity (smoothness), and the orientation of the shadows can all be adjusted to achieve a desired effect (see Figure 2–1e). The price for this automatic sophisticated imagery is paid in terms of added computational time (as usual) and often requires extra memory (RAM). An alternative, which is used by some of the best animation production groups, is creating make-believe shadows from black or semitransparent black polygons that mimic shadows. The cost savings with these simple additional visual cues can be substantial.

A virtual camera is another essential aspect in the visual process. At this point the animator is functioning conceptually as a motion picture camera operator. The same choices as with a real camera are available, such as tilt, pan, dolly, zoom, and focal length. The animator must determine the best angle and select the most appropriate lens for each scene. A program that provides multiple views of the objects is an especially helpful tool here. With four views on the screen simultaneously (front, side, top, and camera's view), the animator/camera operator can readily orient the camera and envision effective viewpoints. This multiview approach is particularly helpful to animators, who must visualize a scene and, at the same time, attempt to anticipate potential problems peculiar to the animation process. Something you'd like to avoid, for example, is objects colliding into—and through—each other. You can see this best with software that provides simultaneous multiple views.

Like a motion picture lighting director, the animator must decide how many lights should be placed on the set and where. Experimentation here can help you to come up with just the right look. Most of the better animation programs provide flexible lighting options such as spotlights and *dark lights,* (the opposite of lights). Dark lights help you control glare, whereas spotlights allow you to cause it—a particularly useful feature to use on the relatively static letters of the flying logo (see Figure 2–1c–e). Dark lights, incidentally, are capabilities not to be found on a Hollywood set.

Adding Movement. Once the image you've created has been infused with a degree of realism, it's time for action. Motion control is another area where the animator needs to think like a combination motion picture director and cinematographer. Fortunately, most animation programs can walk you through this phase; otherwise, it can be confusing and tricky.

Suppose you wanted the letters of the sample logo of Figure 2–1 to appear out of range at first and then to zoom slowly toward the viewer, at the same time angling in for a full frontal view. Your initial task would be to create the first and last frames of the sequence. These are called the *key frames*. The computer can then be assigned to create the frames in between—this *tweening* process is called *interpolation*. As you might imagine, tweening speeds up animation production considerably because it requires the setup of only 2 frames in a sequence that otherwise might have required 20 or more. The computer, doing what it does best, calculates all the intermediary frames automatically.

Now, suppose that after previewing this movement you decide instead that you want the motion of the logo to start from out of range, slowly ease into view, speed up, and finally slow down to a resting position in the foreground. You decide, in other words, that you need to script a nonlinear kind of motion, where the object's motion path is not constant. Most animation software products make provisions for this kind of control of the variability in the motions of objects (including the camera), and allow you to control their speed via interactive graphs. The graph does not necessarily have to be a straight line but can be a curved line—a spline. Its attributes can be readily adjusted so that you get exactly the kind of motion you envision.

Once you've graphed an object's moves, you need to know if you are really getting what you had in mind. How will it look when it's all put together, that is, when it's animated? By virtue of a *preview mode*, most animation programs can create sketchy low-resolution frames (either in wire-frame or fully shaded) so that you can get an on-demand sneak preview of ongoing work. The animation program can accommodate you because it automatically saves working images in speedy RAM, where they can be played back and viewed in real time.

For historical reasons, this motion sampling is often referred to as a *pencil test*. Decades ago—and for the same reason—classic animators would first view their animation in penciled outline form to make sure that the motion had been correctly worked out. At this juncture, if all has proceeded according to plan, it is time to put the motion-scripted images onto video. Surprisingly, this is technologically difficult and adds cost to the animation process in terms of time and money.

Transferring onto Video or Film. An essential and potentially problematic element is the videotape recorder or VTR (also called VCR). Because these are mechanical devices, they have operational nuances that sometimes make them difficult to control. For example, in order to capture a single computer-generated frame onto videotape, the VTR must go backward, stop, accelerate to proper speed, and then initiate recording of the particular frame. For even a short animation, this process is time consuming and hard on both the VTR and the tape.

Luckily, the process can be automated by the use of a VTR controller, which can be purchased either as a plug-in circuit board that uses an available computer slot or as a stand-alone box that communicates with the computer-animation system via a serial port. The catch here is that the software must be able to communicate with the controller and the controller, in turn, must be able to talk back to the VTR. Compatibility is essential.

So what exactly is video? If you can see it on your computer screen, is it recordable on your home VTR? Not likely, unless the graphics board you're using is truly video compatible and has *genlock*, the ability to adjust its signal timing based on the circuitry built into your particular VTR. Unfortunately, without true video compatibility, you are practically guaranteed problems. For example, you may be able to record and play back a nonconforming video signal but not be able to play it back on another machine. Even if you can play it back on another VCR, it may not be editable. Video compatibility problems of this sort are best avoided, as they quickly become formidably complex. The key here is to be sure ahead of time that you are working with compatible equipment. As you might imagine, this is yet another area that benefits from vigilant planning.

Video editing is an integral part of video pre- and post-production. At some point toward the end of production you will need to assemble the various animated sequences recorded on video (and combine them, if necessary, with live-action video sequences). The editing process typically gets very involved and very complicated, so you'll find that taking good notes on your sequences and how they are to be ordered is essential. This documentation process is usually referred to as the edit decision list (EDL). The listing helps you determine where and when the various parts of the animation are to be used in the video production.

Video is not, of course, the only medium available to the animator; you can also put your animation on film. Motion picture film offers outstanding quality in terms of color depth and resolution. The capability comes at a cost, however: Television technology is ubiquitous, and so film is, relatively speaking, an inconvenient viewing medium. Motion pictures can be shown only at facilities that can supply a suitable motion picture projector. There are also penalties on the production side. Film requires more production time (due mainly to the higher resolution) and typically more editing time because the integration of sound is more complicated (compared to video). On the plus side, getting the images onto film is a somewhat easier process. The methodology is facilitated by a special device called a *film recorder*, which automatically exposes the individual images of an animated sequence onto the film. Also, connecting the film recorder to the computer is relatively easy.

Final Words

Animation is a complex business, and so good planning can considerably speed the progress of your undertaking. You will discover that effective planning depends on your ability to predict. Keeping good time and expense records can provide you with a better idea of what to expect when it comes time for the next job. When you know how much time is required to create standard kinds of models and then script their motion, you can estimate costs effectively (as well as have some idea about profitability), and you will be prepared to make accurate estimates of costs on new jobs.

A Low-Cost Shortcut to Video. Although still-frame video animation offers the best quality computer animation, some projects don't necessarily require such high-quality output. Given an IBM-PC/AT ('386 or '486) computer, VGA-type graphics boards are available that provide video recordable outputs. When this capability is combined with a product such as AutoDesk's *3-D Studio*, images can be created on the VGA board and displayed (and recorded) in real time, that is, as the animation is running live from the screen. There are, of course, disadvantages to ready-made animation: Color limitation is a noteworthy one. The main catch here is that the images are typically restricted to something like 256 colors out of a palette of 256,000. This lack of color depth means that shading appears coarse and that lines that run on the diagonal will be somewhat more jagged than they would be when properly rendered. However, when viewed on a standard low-cost VCR, much of this degradation is minor because the image is, in effect, *smoothed* by the inherent quality loss of a standard TV signal.

The benefit of going this route is that you don't need to purchase an expensive VCR or VCR controller—the animation can be recorded directly from the computer exactly as it plays out on the screen. This is a powerful capability. Think about it. There are times when it's far more important to communicate a concept than it is to dazzle associates with eye-popping, feature-film–like special effects. For workaday presentations, such as engineering proof of concept or preliminary architectural viewing, this is clearly the way to go.

This trend toward real-time video recording is even better supported on the latest generation of VGA boards. Many boards, for example, offer 32,768 colors along with video-recordable output (Truevision). Display boards such as these are designed to support AutoDesk's *3-D Studio* directly and also work with AutoDesk's 2-D animation product, *Animator*. Other products, such as *TOPASVGA* from AT&T, provide professional animation capability on everyday MS-DOS–based computers.

In this respect, a daily log can be an invaluable tool when you need to do some trouble-shooting. You will discover soon enough (if you have not already) that in animation production, problems are inevitable. And more often than not, in time of crisis, their solution goes unrecorded.

This is where a daily log can help. Details such as what version of software you are using, compatibility of image file formats, temporary and final disk storage sizes, notes about color drifting on the monitors, and many other details can help significantly when it comes to isolating problems and eliminating them from future production.

With enough practice you'll find that most of the mechanics of the process quickly become second nature and, in practice, flow smoothly. Describing a complex process such as computer animation often has the effect of over-complicating it. Learning to animate on the computer is probably not as daunting an endeavor as it may appear.

What is so wonderful and yet so challenging about 3-D animation is that it still requires so much of its practitioners. The accomplished computer animator has developed a sense of aesthetics and design, a good feel for motion and color, the ability to attend to detail, a tolerance for redundancy, and a good sense of timing. Indeed, computer animation requires not only technical competence but, more importantly, visual and artistic competence. When these Promethean demands are satisfied, the creator is often as rewarded as the audience.

3

Getting Started

The previous chapter provided a fairly concise overview of the animation process—and with good reason. As you set out to purchase a system, you should keep this global viewpoint in mind. At this point, some additional exploration into the process of creating an animation should further sharpen your ability to discern not only your present needs but also your future needs. After all, an effective animated sequence is based on well thought out ideas that have been incorporated into a series of appropriately crafted visuals. To accomplish this complex task you need capable tools that are competently employed in a carefully executed process. The more aware you are of the process and what you want to accomplish in terms of animation, the more likely you are to select hardware and software tools that are suited to your needs. This chapter reviews the crucial planning process so that you can get an idea of what tools you are likely to need at each stage of the animation.

In planning an animation, you need to consider the following:

- the details of the overall process
- developing the story
- integrating sound and editing for final delivery
- the additional hardware and software tools required

Preparation and Planning

Before attempting to create an animation on computer, you should observe the cardinal rule of animation: plan ahead. Animation is so complex, laborious, and time consuming that even the smallest omission can result in massive additional work later on. Without careful planning, repairs may exceed even your most dismal expectations. Omission of even the smallest or seemingly most insignificant of details regarding final editing or video signal levels can result in complete failure of a project or of a vital sequence.

As a simple example, consider this seemingly minor oversight: Two sequences are planned and completed, but no transition has been anticipated to connect them. Because of this poor planning, the animator would probably have to redo both sequences and re-render them in order to accommodate the transition.

The Storyboard

An essential preliminary element in a good animation is a clear description of the story, told in pictures. This way you (and everyone else involved) can get a clear idea of what the animation is about and what it's supposed to look like when it's completed. This sequence of small pictures describing a proposed animation is called a *storyboard*. If the preliminary storyboard is done well, problems regarding lighting and object placement (will one object be covered up by another?) will often surface. Storyboards include details of narration, music, and timing, as well as information about miscellaneous graphic elements.

Storyboard formats vary considerably. A formal kind of animation storyboard can be mapped out on posterboard (about 2 × 3 feet). Cutout pencil drawings show key moments of the animation drawn on black paper. A less expensive and more interactive storyboard uses rows of 3 × 5 inch cards, as shown in Figure 3–1. In this format, each card gives details of the essential features of the animation. These cards can then be laid out or tacked with pushpins onto a corkboard in the proper sequence. This format allows for considerable flexibility when the sequences need to be revised.

Once details of the animation have been storyboarded, you'll find that it's easier to plan for other supporting features, such as timing for

Figure 3–1. *The storyboard acts as a blueprint that all involved can discuss and use as a reference. (Reproduced with permission ALP-Video.)*

music, transitions between scenes, and the placement of required photos or live-action sequences. Based on this level of detail, the next important step can be taken: You need to estimate how long it will take to produce and assemble the animation's component parts.

An issue common to all professions is a particular problem for the animator. You need to be able to predict with some accuracy how long it is going to take to produce a finished product. This is especially important in commercial work, not only because being able to complete your work on a deadline is such a strong asset but also because you need to estimate realistically how much the completed work will cost (and thus what to charge for it). Time requirements depend on many factors. How quickly can you create the models and motion scripts? How long will it take your computer to render the images? How much extra time should you provide in case something goes wrong?

In answering these kinds of questions, you can see that you have to know what your system can and cannot do. Knowing for certain the inherent limits of the system is essential in determining how much time will be required to complete the animation.

Planning Details

While planning the animation you should anticipate the time costs and expenses of such supportive details as editing and special effects. You can often do the editing yourself, especially if it doesn't require use of a complex editing system or necessitate numerous passes. If you plan on using an outside editing facility, however, you should visit them and see exactly what they can and cannot do and at what price. For example, many local TV stations or community cable stations have basic editing facilities, the use of which they may trade for animation work. You need to find out how long it takes to set up at the unfamiliar facility and the availability of personnel and equipment. Will they be able to do the job when you need it done? What kind of special effects capabilities do they have? Are there compatibility problems, and if so can they somehow be easily and inexpensively solved?

Special effects, although usually associated with the fanciful and often bizarre optical effects seen in motion pictures, include many workaday techniques. You will find that many useful special effects are chroma-keyed (this is discussed in Chapter 14). These effects include more than one kind of video transition such as changing the overall color of a scene, etc. If your animation is to be put onto video, you need to find out exactly what special effects systems are available to you. Take the time to visit a first-rate facility (called a post-production facility) and survey their capabilities.

You may also find out that a video production facility may be able to provide complex capabilities, such as the compositing of multiple sequences. Keep in mind that these services will most likely incur some expense, not only in terms of price but also possibly in terms of their effect on overall visual quality. Prices and capabilities vary considerably, so it is important that you know your medium (software and hardware) so that you can ask the detailed questions that will help you to get an accurate estimate of the cost, availability, and time requirements. Finally, as you come to develop some virtuosity in the operation of your system, you may be surprised to find features or capabilities that you can replicate or even improve via your own computer.

The Story

If you want to know what makes for an excellent animation, the answer is simple: an excellent story. A well executed animation of a dull, mixed-up story will convey nothing but haphazard planning and will present the viewer with a confusing and unpleasant task. So what makes for a good story?

Here are a few ideas. A good story tracks well: It has a kind of internal consistency even though it may have surprising elements. Upon reflection, the viewer will see that every part of the story, every sequence and every small detail, contributes to the story and moves it along to its conclusion. Therefore, in a good, well told story there are no unnecessary pieces. A good story is not emotionally, literally, or figuratively monochromatic. It includes contrasts and dramatizes or highlights important characteristics of the animated characters. Elements of anticipation, visual cues, and careful timing direct the viewer's eye, ear, heart, and, ultimately, understanding.

A good story includes some or all of the following elements:

- intrigue
- character development
- anticipation
- surprise
- humor
- appropriate transitions
- effective pacing
- resolution or closure

These basic elements can be used in almost any kind of animated story. The animator also needs to know or learn the language of color and design (especially for applications that do not tell a story, such as flying logos and illustrations). Classic art books on color theory apply equally to computer animation. Knowing that blue backgrounds, for example, are well suited as a backdrop for something bright and dramatic in the foreground is useful information when it comes to composing a scene.

Visual theory can do much to help you illuminate and highlight the essential aspects of your story. When you have learned to manage the images in your animations the way a theater lighting designer or cinematographer would, you have access to potent emotional horsepower. You can harness this power and use lighting and color as artistic allies. You gain control over your medium and can emotionally incite the viewer when it serves your creative purposes.

Much can be learned from expert practitioners. Actively watch TV—it communicates in the visual language of the day. You need to be well versed on the current language of the medium. Your ability to articulate visually is a fundamental and essential skill of your craft. You might also consider keeping a scrapbook of images that effectively convey particular emotions. This reminds you of (and may demonstrate to a potential client) the emotional range and power that can emanate from well crafted images. Lighting a logo from below with red, for example, often connotes fear and looming power.

Finally, you need to understand not only your craft and your material but also the attributes and limits of your intended audience. You should remember that even though typical viewers may not be experts on the subject presented, they are probably visually sophisticated and visually literate. If the audience is to include U.S. viewers, for example, you can assume that they experience several hours of top-level broadcast-quality television per day. They are used to and expect high-quality visual communications. Therefore, if efficient communication is your objective, then visually powerful tools will most likely be a requirement. This is one reason high-quality computer animated sequences are increasingly being seen in unexpected environments and applications. The average viewer is so visually experienced that anything less than state-of-the-art animation is seen as second rate or inferior, and this perception negatively affects the message of the communicator. To a large extent, to echo Marshall McLuhan, "The medium is the message."

Planning Sound

Integrating sound in the animation cannot be overlooked or considered after the fact. Although there are no rules or standards as to when and how to add sound to animation, the issue is important. Luckily, there are many alternatives that originated from cel animation decades ago.

The classical cel animation production process often started with a sound track of the voices and music. The animation was drawn so that the movement matched the timing of the sound. This is how sound synchronization, or *sound-sync*, was (and still is) achieved: It's how audio details such as voice inflections are accurately portrayed and timed visually.

Now, through use of a computer (and established standards for marking time), sound-sync can also be generated after the animation is created. In practice, the way sound-sync is managed depends on the job. If the narration (called voice-over or VO) is most important, the animation is usually set up to follow the timing of the script. Timing diagrams or charts guide the animator to synchronize the pictures with the sound. However, when the animation can be produced first and the sound track can be added later, the sound-sync process is easier to control because you can accurately drive synthesizers and sound effects equipment via the computer. Regardless of what order is chosen, sound synchronization needs to be planned to avoid later "patching and matching."

This brings up the issue of compatibility. Even though the standard time-code processing (and reading) is relatively straightforward, you need to be sure the sound studio's equipment and signal type is compatible with that produced by your system.

Tools of the Trade

Beginners in computer animation are easily overwhelmed. When so much is completely new and unfamiliar, attempting the assembly of a functional system can be a frightening experience. It may or may not help you to know that you're not alone: The whole industry is brand new. You're a technological pioneer. Animation software is constantly being revised and enhanced, and the hardware is changing—mostly for the better. As you assemble the components of your animation system, try to anticipate your overall production process as much as possible. Try to match your personal tastes, ambition, ability, and finances to an existing system while keeping in mind the fact that the system will soon be obsolete.

The components of today's computer animation systems include a daunting array of peripherals, including video manipulation tools and mass storage devices. A sometimes baffling variety of possible combinations enables computer animators to achieve particular artistic aims. However, changes—both good ones and bad—are inevitable. Despite the inevitable array of high-tech equipment that comes with the territory of computer animation, try to keep your goals in mind and not limit yourself to technological solutions to your artistic vision. Regardless of how powerful your computer animation system may be, it is not usually the answer to artistic problems. For example, a super-fast PC may be perfectly suited to rapid rendering, but if you are to be working primarily in 2-D animation, you would be better advised to buy more memory and forego an investment in raw computational power.

Perhaps the most advanced systems are the ones in use at professional computer animation facilities—the ones that are used to create commercial-grade videos. In the best of all possible worlds, you would want to position yourself in the immediate vicinity of such a video production or post-production facility so that video animation work would be frequently available to you and feedback would be immediate. This symbiotic relationship would also put you in a position to use their video equipment at a lower rate in return for providing low-cost animation services to them. Many are surprised to learn that a major part of the expense of animation production is due to the cost of video recorders and other video peripheral support equipment.

You should consider a number of strategies as you equip yourself for animation. It may be more economical for you to get the fastest system available, which is admittedly expensive. Keep in mind that support gear, such as memory, video recorders, and other peripheral devices, will rapidly add to your initial costs. You may want to consider setting up two workstations, which may or may not be networked: One could be used to create the models of the objects, and the other could be dedicated to the rendering of the finished images onto tape. Whatever your setup, you want your professional environment, including all your electronic tools, to be as dependable and predictable as possible.

The pages of magazines are jammed with numerous temptations to violate this principle. The newest and latest gadgets may seduce you into thinking they are must-have items, but they can be far more costly than they appear to be. When you add up the enormous amount of time invested in equipment, as well as in debugging and fine tuning in order to integrate some device into an already existing environment, you may be astounded at the real cost. Therefore, because time is money, evaluate your expenditure of this valuable resource and try to factor it in to the buying equation.

Another critical aspect of the working environment is the quality and capability of the input devices. Some systems (both expensive and low cost) use mice instead of tablets. Some of these systems also provide trackballs for fine control. You need to find out which input device feels most comfortable to you; otherwise you may find that your working environment is less than optimally facilitating your work.

Another aspect of the environment that bears careful consideration are such software features as windows and pop-up and pop-down menus. Some software products allow you to choose display types and whether you want a two-monitor display instead of a single-monitor system. These systems allow you to use both menus and pop-down displays. All of these options are available, and you should weigh carefully their utility so that you can set up the working environment most conducive to your particular working style and production requirement.

Finally, choose your system around good software. A fast computer system is worthless if you can't make it do what you want. Evaluate all the software packages on as many desktop systems as you can, and become as conversant as you can with the myriad of features and qualities that each product highlights. The sooner you ask the detailed questions that relate to your needs, the fewer problems you will have to deal with later—when it may be too late.

Know Your Tools. No system is perfect. Even the most expensive mainframe systems have operational nuances that get in the way of the animation process. In day-to-day operation, even the most sophisticated animation tools can adversely affect a given animation's quality, efficient processing, or a particular animator's creative interaction.

Animation software programs must manage many diverse elements. The software not only controls the hardware but also should be able to assume control of VCRs and other peripheral equipment via the computer hardware. In the best case, all hardware and software elements function smoothly together and speak the same language. In other words, they are operationally compatible. Indeed, if one element is not communicating properly with another, the animation may be catatonically suspended with little or no notice to the user.

You will discover that every system has its limitations. You will find limits in terms of resolution, shadow generation, interfacing (with different pieces of equipment), and motion control. You will certainly run into limits when creating rounded objects. Only when these limitations are kept clearly in mind can the process of creating an animation be realistically planned out and executed without unfortunate surprises.

What if you aren't sure about your system's limits? The best way to find out is to experiment with any questionable capability or function: Check out anything you aren't sure of. You need to eradicate unwarranted assumptions you (or the developer of your software) have made about your system's capabilities. If in doubt, test it out. You could save yourself an astounding amount of time, bewilderment, and energy later on. This book discusses hardware and software issues in depth in later chapters.

The Computer Setup

4

One trend in the realm of computer animation is the constantly increasing power of the hardware. Capabilities and features that were available only in the rarified domains of mainframes and high-powered graphics workstations are now available to the rest of us for use on ordinary personal computers. In fact, it is now difficult to differentiate clearly the attributes that define a high-end desktop personal computer (a PC) from those of a low-end graphics workstation. The lines of distinction increasingly blur. Indeed, some 3-D animation programs that were originally written for mainframe platforms can now be run on desktop systems.

The pace of technological improvements has been so rapid that graphics workstation vendors are now manufacturing plug-in circuit boards for PCs that possess all the power and capability of full-blown workstations. There is, however, a notable downside to all this heady innovation. Although improvements continue to offer greater capability, in practice animators function as the computer industry's technological guinea pigs. Indeed, over the past few years most computer animators have been, in effect, unpaid beta testers (those who try newly made but unreleased products), as they pioneer day-to-day use of the new hardware and software and test the various products' overall compatibility in individual desktop systems. [In the appendix, we provide a list of functionally compatible animation systems.]

Another effect of increased power is evident in the way animation is used on these lower cost systems. Due to their increased RAM, greater hard disk capacities, and faster video access times, personal computers are being used as interactive video editing centers, video control systems, real-time video animation platforms, and, in some cases, as professional 3-D and 2-D video broadcast animation stations.

Currently, the most popular PC platforms suitable for animation tasks are the Apple Macintosh, the IBM PC/ATs, and the Commodore Amiga. Reflecting their technological roots as well as a particular philosophy of design and operation, these computer platforms radiate their own unique (and peculiar) personalities. Naturally, each platform capitalizes on the advantages of its exclusive technology and downplays or attempts to ameliorate its technical drawbacks. Of course, each has its own devoted following. It's not that any one platform is superior to the others: rather, each marshalls a complex set of technological attributes that serve to define its targeted user's needs.

Due to the built-in adaptability inherent in IBM's choice of an open architecture, rapid synergistic improvements to the platform continue to spin off from third-party developers. Indeed, the PC/AT platform has evolved to become the first low-cost computer to provide high-quality, broadcast-grade 3-D animation. Third-party manufacturers of graphics boards have supported and significantly enhanced the capabilities of the system. Likewise, advanced video control boards have provided the platform with a straightforward path to video broadcast facilities. Perhaps the main reason for the popularity of MS-DOS computers for high-quality computer animation is because the peripheral components of the system are improving so rapidly that animators can be increasingly competitive while extending their creative capability.

Although the number of animation products designed and marketed for the IBM PC/AT domain are few but excellent, the sheer number of animation products for the Mac is remarkable and still increasing. A couple of the Mac's unique features have led to this plethora of products. The Mac's user interface is standard; and this fact, coupled with a bit-mapped screen standard, facilitates movement from program to program. Users can switch to a paint program and then to an animation program without having to deal with mental conversions and the overhead of dissimilar or even contrasting graphical syntax or file format.

The Mac is increasingly being used as a graphics do-all. A host of different kinds of paint and animation programs are available and in wide use, and at the same time the Mac is often used to serve as a foundation for other allied processes (for example, as an editing controller).

The Amiga is outstanding in many respects, but what makes the Amiga unique, particularly in the context of computer animation and graphics, is that the machine was intended and designed for video from the very beginning. Several animation products render complex 3-D images and, after the animation phase is complete, either transfer the animation onto a video recorder or save it in RAM and play it back live in real time. From there it can be recorded directly on any inexpensive VCR. Amiga software products typically take advantage of the desktop's powerful processor for multitasking and for internal sound generation. For example, NewTek's Video Toaster is an Amiga hardware peripheral that transforms the computer into a video special effects system, a video switcher, a video capture device, a character generator, and a video mixer.

Devoted users of these three popular systems develop biases that, for emotional or logical reasons, tend to support their point of view. IBM-PC users, for example, cite lower system costs and greater upgradability than Mac systems. Mac users, in turn, point to the continuity of the user interface as they move from product to product as well as to the file standard that facilitates ease of use. Amiga users point to the desktop's low cost and excellent video integration as primary advantages. Selecting a platform is a complex decision; however, there are guidelines. Perhaps the most important one is to consider all aspects of a system, in particular the kind, quality, and cost of the software that is available for it. Indeed, astute prospective buyers begin by making their software choices first and then purchasing hardware components that support the software.

Integrating Your Own System

There are more than a few good reasons to select any one of the three computer platforms already discussed. The choice should be made on the basis of the particular needs and resources of the user. Many factors need to be plugged in to the buying equation: How much memory will be required to run your applications? What about backup capacity? What resolution quality is acceptable to you? How about system speed (how fast is fast)? Will you require multiple stations? What about an upgrade path?

These are only some of the considerations that might affect your selection of system hardware. Because a wide range of hardware options are available to the animator, a functional system represents a series of decisions, often technical in nature. Considerations such as video compatibility and the availability of task-appropriate software must be carefully evaluated. After all, a PC-based system must be set up so that it can efficiently produce the kind of animation you envision, and it must do so cost effectively.

Whatever animation system you choose, you will find limitations and strengths in the ways the parts of the system interact. For example, one system might be easily dedicated to the labor-intensive process of creating a model, whereas another system might be optimized for the automatic, yet time-consuming, process of putting sequences of images onto videotape.

This brings up the often asked and not easily answered question: How does a prospective buyer decipher a manufacturer's cryptic (and often misleading) spec sheet? The short answer to this question is, in a word, carefully. This chapter is, in effect, the long answer to the question. The better informed you are with regard to your own needs, the more likely you are to get what you need the first time out.

If money were no object, a highly evolved and specialized graphics workstation-based computer system dedicated to your target animation needs might be your best choice—as long as your animation focus doesn't change. Then again, such a system is likely to be obsolete within a short time. What's more, with software graphics workstations choices are more limited and upgradability is often more expensive.

For starters, it's a good idea to check out the operational context. A desktop PC-based system could be part of a larger animation environment. For example, a low-cost Amiga might be used for model development in a system in which a more expensive graphics workstation system is used to do the rendering. Or a PC might be used to capture video images from an existing video sequence and to composite a video image onto a computer generated scene. Such dedicated uses optimize the capabilities of a PC and at the same time free up other equipment for such computationally intensive tasks as rendering and compositing.

Overall System Speed

The overall performance of a system is based on a number of interwoven attributes, such as central processing unit (CPU) clock speed, bus speed, and bus data width. Each component must be well matched to related components. For example, if a graphics board or a memory/graphics board is placed in an 8-bit slot when it is designed for optimal operation in a 16-bit slot, throughput will be halved.

Other factors also affect system speed. Even though computational speed is essential for 3-D animation because of its extensive use of floating-point math operations (calculations such as 13.59 divided by 0.12), this kind of processing may account for about only 15% of an image's total rendering time. Likewise, if the system's hard disk is used extensively for swapping digital images or data, it too needs to be fast enough so that it does not bottleneck data flow and slow down other dependent processes. System random access memory (RAM) can pose a problem too. When extra RAM or RAM extension boards are limited either by the board's throughput or by RAM speed (measured in nanoseconds), a system's functional throughput can dramatically slow down despite the host processor's relatively fast operation.

Some computers have incorporated optimization schemes that help boost effective overall speed. Compaq's MS-DOS computers, for example, utilize a proprietary hard disk interface that is faster than the standard interface. This means that files or images that are called up often during an animation process will be loaded with considerably more dispatch, resulting in noticeably faster disk-intensive operations.

Your Host: The Desktop Computer

The *host computer,* as the term implies, acts as a hardware platform that provides the environment within which the peripheral components operate to extend or enhance the host's capabilities. Because the newer PCs are typically engineered to accept second-party-engineered peripherals, the system's architecture is said to be *open*. The architecture of the early Macintosh computers was *closed:* They accepted second-party peripherals only via elaborate re-engineering schemes.

PC architectures vary considerably. Clock speed, addressability, word size, throughput, and expandability all differ from one system to another and can materially affect (either enhance or limit) the capabilities of the animator. Although system speed is perhaps the most important factor in overall desktop performance, peripherals can play a critical role in a system's total performance.

Because the PC platforms we are focusing on function primarily as hosts for a broad range of second-party and even third-party peripherals, a prospective buyer needs to look critically at the availability and cost of add-on components. These components are usually pivotal to the overall effectiveness of the system. Figure 4–1 shows a typical setup of a host computer with its peripherals.

Indeed, peripherals are so important to the operation of the host that the following discussion concentrates primarily on the peripherals and how they interrelate with one another and with the host. Available peripherals are so numerous and functionally so important that they can rapidly eclipse the cost of the host PC to which they are connected. Furthermore, the ability to accept an almost endless array of hardware adaptations makes for an immensely versatile—and updateable—system. Thus, it's important to evaluate the costs of all needed peripherals in order to come up with an accurate assessment of a functional system's actual cost.

All PCs included in this discussion (the IBM-AT 286, 386, and 486, the Amiga, and the Macintosh) boast open architectures. Because of this it's not necessary to dwell on the specific capabilities of a particular host or to focus on the advantages or disadvantages of one platform with regard to another.

The Extras: Peripheral Components

Some peripheral components are so complex and so integral to the animation system that they are discussed as separate topics and have chapters of their own. These include graphics board(s), monitors (including a discussion of RGB graphics), and composite video technology. The other major peripheral contributors to desktop animation are discussed in this section. A host computer can incorporate any of the

Figure 4–1. *A computer animation system should be set up for reliable, efficient, and easy-to-use daily operation. Shown in the computer/VTR box is a '486-33 Micronics system with a Micropolis 480-megabyte hard disk for 3-D animation; a JVC 7500A S-VHS single-frame recorder; a Wyse '386-25 for 2-D animation, paint, film-recorder transmission, and mass storage; and a Panasonic editing VHS deck. The system is completely integrated for continuous automatic operation.*

following hardware components, each of which can significantly enhance a system's overall capability:

- system memory enhancements
- math coprocessers
- video and film recorder interfaces
- manual input devices
- mass storage

The following subsections give a closer look at these performance-boosting peripherals so that you can see how they affect the animator's desktop.

System Memory Enhancements. Computer graphics applications are notoriously memory hungry. These days, as programs grow in richness and complexity, it seems you can never have enough RAM. Newer software programs increasingly take advantage of (indeed, they demand) extra RAM. The programs themselves are large and the amount of data to be manipulated is even larger. Capturing an image and applying it to a rendered surface is most easily accomplished while the digital image is stored in RAM (along with the 3-D model). The alternative is to access the image via the hard disk every time one of its elements is to be placed on the model. The difference in overall performance is dramatic. When adding more RAM to your computer, make certain that it is designed to perform at the system's optimal speed (measured in either megahertz or nanoseconds). Emulation of RAM, by the way, runs slower than does the CPU's RAM.

Math Coprocessors. If your target application requires a lot of rendering (as does 3-D animation), the host computer must provide the computational power for timely rendering. If performance is sluggish, a math coprocesser chip (or board) should be considered. It should be noted here, however, that the addition of a fast math coprocesser won't (necessarily) increase performance of 2-D or paint animation because these processes do not often need complex math calculations.

In computationally intensive applications, a math coprocessor can make a huge difference: Rendering a 3-D image can take as much as hundreds of times longer without one. Therefore, as you might expect, the speed of the chip is important: the faster the better. Many animation software packages require the addition of a coprocessor and will not operate properly without one. In the case of '486 computers, a math coprocessor has been integrated into the host's main processor chip. Although competitive second-source coprocessor chips are available, the advantage in terms of performance is usually negligible.

An alternative to a math coprocessor is a plug-in accelerator board. Such a board takes on the whole process of rendering and maximizes the entire process with performance superior to that of most math coprocessors. Accelerator boards utilize a specialized processor in conjunction with several megabytes of fast RAM and are thus expensive. When rendering time is critical, however (such as in large-scale commercial animation production), an accelerator board can reduce total animation time by a factor of four and thus pay for itself in a matter of days.

Video and Film Recorder Interfaces. Another essential component is the interface between the VTR (or film recorder) and the computer. Although this topic is covered in greater detail in subsequent chapters, it is introduced here. The interface is accomplished either by way of a special plug-in circuit board or via the host's serial port. This interfacing is critical to the animator because recording devices must be synchronized to accept a sequence of images from the host. Timing is critical: on-board intelligence coordinates communications and operations between host and recorder.

When frame accuracy of the video tape is not important (for example, when an animation is played directly from the computer), low-cost VTRs can be operated via the serial port. This simplifies the connection between the computer and the VTR because the only interfacing needed is a serial cable. Software can control the operation of the VCR directly.

Manual Input Devices. The choice of a graphics input device has considerable subjective elements. Whatever you use—graphics tablet, mouse, trackball, or a combination—the graphics input devices must be suited to your experience and ability. A key factor in the selection should be the type of input required, as determined by the software you regularly use. However, regardless of the input device, remember that you'll use it a lot and so you need to be sure it's well made and can stand up to long term use and even occasional abuse. The following paragraphs summarize the dominant attributes of the various input devices currently on the market.

Graphics tablets enable exact control of data input, especially for such tasks as model creation. For larger models, which require the establishment of many contiguous points, the large table-sized tablets can provide tens of thousands of points of object size. Two types of locators are generally used with tablets, the stylus and the puck. A *stylus* is good for drawinglike functionality, whereas a *puck*, which looks like a thin mouse with a transparent crosshair, allows for accurate placement of points. Pucks, like mice, include option buttons to signal a routine relative to some point in the graphic. Some styluses are sensitive to pressure variation. They are used mostly in support of paint systems, where pressure variance changes the color density or airbrush width. Such devices are, however, limited for use in addressing 3-D positions as required by a modeler, despite the fact that animators often switch from a 3-D program to a 2-D paint program (for touching up details in a picture, for example). Still, having the extra capability of the pressure-sensitive stylus for use with a paint system a useful.

A Precaution. Because most tablets rely on sending and receiving magnetic fields, it isn't a good idea to put floppy disks on the work surface unless you know that the tablet is a nonmagnetic type (such as Kurta).

Mice are inexpensive, and most people find them fairly easy to use. However, despite their obvious usefulness, they are difficult to control in certain kinds of applications. For example, you'll find that using a mouse for extended periods of time requires continual micro-adjustments of the elbow, which can be tiring and frustrating. You'll also find that movement control can be especially difficult when you are creating objects that require fine detailing.

Trackballs have been tried in many configurations but haven't caught on as a staple in the world of hand control. However, when used in combination with another device, such as a mouse, a trackball can be a decided convenience because it offers wide range of scale and is stable after interaction. This means that your position on the screen is retained after you take your hand away from the device.

All of the input devices discussed so far have been designed to control digital objects in two dimensions. However, when it comes to fashioning complex 3-D models, special 3-D input devices can be extremely handy. Creating a 3-D model requires simultaneous manipulation of an extra dimension and is best managed with specialized input devices. When a model of a 3-D object is to be created via computer, key dimensional points of the object must be selected to map the object digitally. Radically different technological strategies are employed to do this. The 3-D input device may generate its data by using sound waves, or it may rely on a gimbaled arm to locate explicit 3-D points and then send them to the computer. The software then assembles the key points and uses them to construct a set of connected polygons. Both kinds of systems have operational limits, the most obvious of which is size. If any dimension of a target object exceeds the 3-D digitizer's size, the object will be, at least in part, completely off-scale and usually will require extensive data editing. These devices are not free of problems, but they make the process of translating a physically complex real object into 3-D data relatively painless.

Mass Storage. Computer graphics applications can get positively gargantuan when it comes to the storage of images. That's why almost all of today's serious computer systems for animation include integral hard disk storage as well as additional *mass storage*. You can hardly have too much memory. Indeed, when it comes to mass storage, you should acquire as much capacity as you can. You may be surprised at how soon you fill it up.

The usual accumulation of software programs, models, scripts, and other data should be accommodated on a large-capacity hard disk with a fast access time. Access time is important because animation programs often require repeated access to the drive. Additionally, the hard disk is used for creating and/or storing images, and for managing large temporary internal data arrays. These uses all necessitate repeated disk access. Slow disk access can, in effect, degrade apparent software performance to the point where it is not only noticeable but also frustrating. Indeed, rendering performance may seem to be primarily a function of CPU speed, but the throughput of the hard disk can drastically impede the performance of a long animation. This is most apparent with 2-D animation software that plays back sequences of images stored on the hard disk.

Removable Mass Storage. Although a hard disk is the usual primary repository of stored data, secondary sources are equally important in an animation system. Every animator needs to address the issue of backup and should consider backing up onto removeable media. After all, hard disks can be operated on the sensible assumption that they will, at some point, fail. The mean time between failure (MTBF) for most hard drives is somewhere between 20,000 and 60,000 hours. There are 8,760 hours in a year: You figure it out.

Popular types of backup media include floppy disks, streaming tapes, digital audiotapes, 8-mm cassettes, digital storage on videotape, optical write once read many disks, and erasable optical disks. The market for backup equipment is extremely competitive: Prices, capacities, standards, backup speeds, and reliability vary widely.

Animation model and motion files consume a great deal of memory. Additionally, image files, which you will most likely want to store in an image bank or library for use in backgrounds or for demonstration purposes, can take up one megabyte or more for each image. Because animation production is labor-intensive—often consuming days or weeks of creative effort—the importance of redundant data files is undeniable. If you don't have backup files, you have, in effect, all your digital eggs in one electronic basket.

If you have the capability, you should regularly back up your files onto removeable media and store the media in a safe and secure location. Portability is an important feature. Portable image data can readily be delivered to an outside company, where it can be read and put on videotape. Of course, your media and format must be compatible with those in use at the production facility (usually necessitating prior verification).

There are two ways to store data: Your files can be distributed as a serial stream of data, or they can be arranged randomly throughout the storage medium. Storage on *hard disks* and *floppy disks* is based on the random access approach. When you request a file, the unit goes to that particular file and retrieves it immediately. With serial storage, however (such as on tapes), all of the storage medium must be searched until a designated file is found.

The problem with hard disks is that they are expensive and not designed to be removable. The problem with floppy disks is that their capacity is marginal compared to the amount required for a typical backup. For that reason, serial devices are used when you need to back up several megabytes or more.

Figure 4–2. *This 150-megabyte streaming tape cartridge is a popular type of mass storage for use in backing up files.*

Serial storage, such as *streaming tape,* or streamers (see Figure 4–2), is a popular solution and is used mostly for very large singular files, such as high-resolution images (a 2k × 2k uncompressed image may be as large as 100 megabytes) or sequential strings of images, or for daily backup. The main advantage of this relatively inconvenient and slow mode of storage and retrieval is low cost. Sizes vary from 44 megabytes to over 150 megabytes. Backing up a low-cost 150-megabyte hard disk each day takes about a half hour. The tapes cost between $20 and $50 and are cost effective in terms of dollars per megabyte, not to mention the insurance factor.

Other forms of serial backup storage include *digital audio tapes* (DAT) and *8-mm cassettes.* These media offer greater capacity than do streamers—often more than 1 gigabyte. Although they are relatively fast in terms of data throughput, when it comes to retrieval you must review the whole tape to recover a discrete piece of data.

Another cost-effective solution to this level of mass storage are *write once read many* (WORM) disks, which can store over 1 gigabyte of information on a single optical disk similar to an audio CD (see Figure 4–3). These devices operate on the principle of random access, so they can store and retrieve data immediately on request. Their operation is similar to that of a hard disk drive; however, their data throughput time is slower by more than ten times. WORM drives are also not erasable. These disadvantages are more than compensated for by the fact that they can store much larger amounts of data and their cost per byte of storage is decidedly favorable.

Erasable optical disks are yet another option. As for WORMs, their formats vary widely, so it is unwise to assume that one system is compatible with another (a potential problem when upgrading to a newer system, even if it originated from the same manufacturer). You can also run into compatibility problems when the optical medium is to be used by an outside service bureau. However, despite these drawbacks

Figure 4–3. *This 1.2-gigabyte optical disk, a stand-alone device, is used for saving large numbers of images for textures, maps and backgrounds.*

(which include slow access times) the convenience of being able to back up massive amounts of information on a removeable disk is undeniable.

Compatibility Issues

Compatibility of components is, of necessity, a major concern for the computer animator and perhaps one of the most serious drawbacks to a PC-based desktop animation system. In a typical system, a number of circuit boards and other peripheral components will be integrated with the host, and so communications among the various components can become a problem. Each peripheral must, in effect, be able to communicate fluently in the language of the host. The potential for incompatibilities is enormous; for example, graphics boards may not operate at the same plug-in bus speed as the host computer, memory chips may be too slow, or the digital film recorder may utilize an incompatible format.

Unfortunately, standards don't often survive the test of time—they are constantly evolving in a kind of technological drift. As a result of constant competitive forces coupled with incessant technological innovations and improvements, popular standards have a half-life of little over a year or two. Because of this, questions of compatibility continually arise. Unfortunately, the ultimate test must be based on hands-on trial and error. Professed adherence to a standard is often not worth the paper it's written on, although its a good place for you to start the weeding out process. Manufacturers can usually provide lists of systems known to work together and those that have problems, so it is advisable to use this information first.

The issue of compatability applies also to software. For example, data files, such as imported image files, must be completely compatible with the software accessing them, right down to the particular version number; otherwise, wild and unpredictable results can be expected.

Some software programs, for example, create data files that may be incompatible with previous versions of the same program. As you might expect, compatibility with other software is even less likely. The issue of video compatibility can be even more troublesome, as you will see when you reach Chapter 13 which covers video interfacing.

Another compatibility issue arises out of the sheer number of devices that can be connected to a host computer system. A tablet, trackball, film recorder, modem, and VTR controller may each require a serial port. Even if there are enough serial ports to go along (IBM PCs, for example, are typically equipped with two), configuring the ports without conflict may be a difficult and time-consuming operation in itself.

Upgrading

Given the current flux of today's computer technology, changes to your system are almost inevitable. Sooner or later you are going to want to upgrade some of your system's peripheral components. Despite the well founded fear of introducing new incompatibilities, new products that can make you more productive are undeniably tempting. When you are about to succumb, you need to remember that verifiable compatibility should continue to guide your decision. Ask yourself whether the new software or hardware component will work in the context of your system.

As you shop for the device that allows you to enhance or in some way augment your animation capabilities, inevitable questions will arise. If at all practical, try to answer your questions about the addition by actually trying it out on your system. You'll find that, on paper, almost everything will work. Whether it will work in your system is a different matter. For example, suppose you wanted to upgrade your graphics board. Will your digital images remain compatible? What about the monitor you use: Is it compatible with the new board? Does the new board use the same cable connectors? Suppose instead you decide to purchase a new monitor. Will the aspect ratio be the same as the ratio displayed by your old monitor? How about colors? Will the colors of existing work look the same (or even close) on the new monitor? Regardless of what new element you introduce into an established system, you will usually have to make adjustments. Keep in mind that the usefulness of a particular scheme or processor ultimately depends on your animation requirements, your plan for upgradability, your budget, and the overall capability you desire.

A complete animation system is a remarkably complex amalgam of interrelated devices controlled by software and hardware. The introduction of almost any new variable into such a complex technological balance can have surprising consequences, to say the least. Do your homework, and whenever possible test each addition in an environment similar (or identical) to your own.

5

The Graphics Display Unit

If your host computer has no dedicated high-resolution graphics output suitable for animation display (which is usually the case), you must select a peripheral graphics board and a graphics monitor. These peripherals are perhaps the second most important attribute of an animation system (the first being the host computer). On some platforms, for example, the Amiga, you may opt for additional graphics capability, such as a high-resolution graphics board, so that you can access a greater range of colors. This chapter takes a close look at some of the features that a graphics board can provide you.

For starters, it's interesting to note that some dedicated graphics display circuit boards provide greater computational capability than the computer system they are designed to support. These high-powered hardware modules usually include specially written software to take advantage of the board's features or capabilities. In practice, this kind of setup allows most host functions to execute normally on the PC while graphics-intensive functions are performed directly via the graphics display board.

To evaluate a graphics board, you need to zero in on its primary functions:

- compatibility with software
- resolution
- color output
- speed
- video interfacing
- special features

Regardless of the special features boasted by a particular graphics unit, you need to keep in mind the animation process you are targeting as you research the marketplace. You also need to be aware of the available software that is compatible with the board. For example, if video resolution is a primary concern, the selection of a board offering resolution higher than video would probably not justify the extra cost. If 3-D modeling is an area you wish to pursue, you need to be sure that acceptable software is available to support the modeling tasks.

Boards designed to plug into computers of the IBM-PC/AT type include those configured for dedicated video use as well as those set up specifically to handle mixed computer and video. Dedicated video boards, such as Truevision's series of graphics boards, concentrate on the provision of such graphics functions as video capture and display. When these types of graphics boards are used, another graphics adapter board is needed to manage text displayed in conventional graphics formats such as CGA, EGA, and VGA.

Mixed-function boards (such as video/VGA boards that display conventional VGA graphics for video output) display conventional computer text in addition to outputting recordable video. These kinds of graphic display adapters can automatically switch resolution when appropriate to support graphics functions. Your choice of software will do much to determine what type of graphics board and monitor configuration is optimal for the way you work.

Graphic board options (see Table 5–1) for the Macintosh offer similar kinds of choices. You can opt for a single, standard monitor that displays text and gives you a window of your animation application. Alternatively, you can use a two-monitor configuration in which a dedicated graphics board provides video input and output and the application software runs on the regular Mac screen. Whether you use one screen or two screens with an additional graphics board is entirely a matter of individual preference.

Table 5–1. Specifications of Computer Graphics Displays

System	*Format*	*Resolution*
NTSC Composite	Encoded	480 lines
Analog RGB	Analog	480 × (X)
IBM PC, XT, AT, PS/2:		
Monochrome	Digital	720 × 348
CGA	Digital	240 × 320
EGA	Digital	480 × 640
Super-VGA	Analog	600 × 800
Super-VGA	Analog	768 × 1024
Apple Macintosh:		
Hi-Res	Analog	1172 × 810, 1024 × 768, 1152 × 870
Medium-Res	Digital	832 × 624

Note: Specific manufacturers of boards often provide different specifications. Furthermore, all specifications change dramatically with time.

Compatibility with Software

One of the benefits of a rapidly advancing field is the availability of continually improving versions of software products. This constant innovation is desirable—until you run into problems. You may even have problems with the latest board and the latest software. If you don't want to take up half your time pioneering solutions to buggy software, it's a good idea to stay with tried and true versions of popular software. Having the latest and the greatest is, in terms of software selection, risky business. Select software that you know will work in your present system.

Other compatibility issues include the possibility that a clone board might lack certain features that preclude the normal operation of certain software applications. You may also find that other software you use (such as a paint program) may not be compatible with your clone board. Try to confirm before the purchase that the board you select is completely compatible with the software tools you use.

Resolution

A graphics board's resolution specifications may be confusing at first glance. Why, for example, should a graphics board provide many different resolutions? The reason is that in order to remain competitive, manufacturers need to support numerous standards. Only rarely does this benefit the user. The result, however, is greater graphics capability and adaptability.

As a resolution benchmark, consider the resolution of the typical home VCR: about 240 vertical lines. This may sound limited, but the image appears acceptable because the specification allows for a vast range of colors. Likewise, a graphics board that offers only mediocre resolution but compensates by supporting a broad color range can produce an acceptable display. This phenomenon demonstrates a key concept of digital graphic displays: The features of graphics boards interrelate to produce visual output. Thus, when evaluating a board, you should attend to the overall desired output quality rather than depending on "specsmanship."

Remember that to use a product's specifications as anything more than a rough initial guideline is risky business. Manufacturers, caught in the game of competitive numbers, may omit key facts or qualifications regarding the features or capabilities of their product. For example, a company producing graphics adapters might rightfully claim support for a particular high-resolution monitor: yet, nowhere are you informed that the board is inoperable at other resolutions (such as low resolution video). Caveat emptor.

When evaluating a particular board's resolution specifications, you should focus on your own animation needs. For example, if you are primarily concerned with the creation of video-grade animation, a resolution greater than video (roughly about 480 vertical scan lines by 512 horizontal lines) would not be useful to you—unless, that is, you are creating complex computer aided design (CAD)-based architectural building models where a high-resolution display (1024 by 768) would clearly be beneficial. In this kind of situation, access to multiple resolutions could be very helpful. You could build your detailed model with the clearest high-resolution monitor and then switch (via software and plugs) to a low-resolution TV-grade video monitor to evaluate and preview the finished look. However, whatever the board's claimed specifications, be certain that the software you want to use supports the board and it's various resolutions.

Incidentally, whatever resolution or resolutions the board you choose supports, you should select a monitor that is compatible with your graphics adapter as well as with your intended application needs. We'll discuss monitors in considerable detail in Chapter 6.

Does high resolution means faster image display? Not necessarily. With high-resolution boards, the greater the resolution, the more addressable pixels and, hence, the more RAM necessary for temporary storage of an image. Thus, high-resolution boards may display an image more slowly than would a system with lower resolution. That's one reason an Amiga can display animation sequences in real time from its RAM-based memory, but only when running in low-resolution mode.

The resolution limitation of a board doesn't necessarily mean that a resolution above the board's specs is unattainable. Resolution-independent software can generate high-resolution data and output it to another device, such as a digital film recorder. Thus, the preview images might appear to be of poor quality (as far as resolution is concerned) on screen and yet, because of the software, images of very high resolution can be produced and viewed on film.

This important concept applies to still images as well. A low-resolution model can often be created in high resolution for slide-making purposes. Even though you cannot display a high-resolution 4000 line image on your monitor (monitors with this capability are not yet available), a high-resolution digital file can be ported to a digital film recorder, which can then process the high-resolution image onto slide film.

Color Output

A graphics board's actual color output capability is often difficult to ascertain and therefore requires careful evaluation. One of the most common areas of confusion is enumeration of color capability. Two numbers are taken into account here: You want to know how many colors are simultaneously displayable and how many colors are potentially displayable in reference to the board's palette.

Before going any further, we need to take a look at the two most common strategies manufacturers use to display colors on graphics boards. One method of color display provides a memory location for each position of the available resolution for a particular color value (including black and white). Thus, if a certain position is filled with an *on* bit for a particular location, that position (pixel) on the display screen will be on. This is described in terms of planes. A *plane* is like a pane of glass with grids depicting where memory locations exist. Each memory location corresponds to the ability to display a particular color. With color-based graphics boards, each plane can provide an additional multiple of colors—that is, 2 planes of memory equals 4 colors; 3 planes equals 8 colors; 4 planes equals 16 colors; and so on (see Figure 5–1). It then follows that 15 planes (2^{15}) provide 32,768 colors. In other words, a system providing 15 bits per pixel is able to display 32,768 colors.

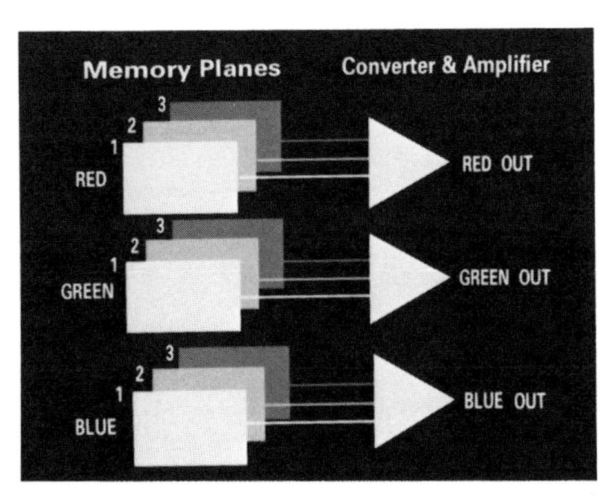

Figure 5–1. *The number of color planes determines how many colors are potentially displayable on the graphics board. Here, the total of 9 planes provides 512 colors.*

If a graphics board can display 32,768 fixed colors (15 bits per pixel, or 5 bits for red, 5 bits for green, and 5 bits for blue), you might conclude that you have adequate color capability; this is not always the case. If, in the course of an animation, a scene is set up to shift gradually and completely from pure black to deep royal blue, only 32 shades of blue (2^5) are available for display. The transition from black to blue will appear ragged, far from the smooth effect you were looking for. What's more, when a solid green line is rotated in a scene, it must be smoothed—*anti-aliased*—so that the edges do not appear jagged. However, if not enough hues of green are available, the line will appear as a coarsely shaded moving succession of jagged dots roving about the edges of a green object.

The second method of color generation does not use fixed memory locations for color display. Instead, a *color look-up table* (CLUT) is used (see Table 5–2). In this scheme, a particular location on the screen (called a pixel) can be assigned any color. A CLUT-based display offers certain advantages to the animator. For example, simple color animation tricks, such as color cycling, can be generated easily. Any color assigned an image can be gradually reassigned to another color value. In contrast, fixed color systems (the method described earlier) are incapable of this kind of flexibility because their memory slots are preset to a particular color value and cannot be changed.

Boards designed to use CLUTs can be evaluated on the basis of their input/output size. The input count describes the number of bit planes in the CLUT. The output count describes the number of variations that each bit plane possesses. A typical value might be 8 bit planes possessing 256 values for each plane. This 8 × 24 color map translates into 256 colors that are displayable at one time (2^8) and a palette (potential) of

Table 5–2. The Color Look-up Table

The total number of displayable colors of a graphics board is based on the number of color look-up planes. This chart translates the number of planes to the number of colors available.

Planes	*Levels*	*Available Colors*
1	2	8
2	4	64
3	8	512
4	16	4096
5	32	32768
6	64	262144
7	128	2097152
8	256	16777216

16,777,216 (2^{24}) colors. Full (true) color is built up on a 24 × 24 color map that can also display 16 million colors simultaneously.

So why the distinction between the two methods of creating colors? The answer becomes obvious when you start viewing your animated sequences. You'll discover that even slight variations in color become noticeable and objectionable when viewed as an animation.

If your graphics board utilizes a large CLUT, say 16 million colors, but output is limited to the display of only 32,768 simultaneous colors, you'll find out that you actually have a greater range of color choices that can be used for techniques such as anti-aliasing. Indeed, you'll find that overall image quality is superior to that obtained by a comparable board using a fixed color table.

Some programs attempt to compensate for the artifacts produced by graphics boards that are limited by fixed color tables. Software routines can attempt to offset the poor image quality by a process called *dithering*, in which random colors are scattered during color transitions so that images are somewhat massaged and the gradation quality is apparently improved.

What does all this means to you as an animator? The answer is that you must evaluate the color capability of a board you intend to use and at the same time you should carefully review your own production requirements. Does the level of color quality suit your needs? On TV-grade video, for example, a seemingly limited 15 bits per pixel (32,768 colors) is not discernibly different from so-called true color, or 24 bits per pixel (16.7 million colors).

Speed

The speed (measured in megahertz) of a graphics board relates roughly to the time required to display pixels. Because advertised speed specifications are notoriously unreliable, speed ratings should be used only as guidelines. Hands-on performance testing is the only dependable way to ascertain the workaday speed of a board, and measurement of the real performance of a board is tricky and can get complicated in a hurry.

One way to measure a board's output speed is based on its ability to display pixels, lines, and polygons. Another way is to measure the time it takes to fill a complex (concave) polygon. However, you should be advised that all tests are biased by the measurement method selected. What's more, a manufacturer may claim very high pixel display rates while failing to mention the fact that an inherent bottleneck exists in communication with the host computer.

Graphics boards can be said to be either smart or dumb. That is, some boards incorporate on-board intelligence with a dedicated microprocessor and thus have the ability to interpret incoming information. Some boards are designed simply to facilitate speedy display of the pixels on the monitor. The speed of a particular graphics board needs to be understood in the context of the overall architecture of the system and in light of your particular animation application.

In applications in which CAD models of buildings or construction parts are being created, advanced CAD programs can take advantage of a feature offered on some boards called a *display list* option. A board with this capability can store the CAD data in its memory, facilitating graphics computation. When you require zooming in on the model, for example, the host computer does not have to do the calculations because the board, which has been designed to optimize this function, executes the zoom rapidly on its own, independent of the host. If your application involves the translation of CAD-oriented models into animated sequences, consider a board that optimizes output processing with a display list capability.

Perhaps the best solution for most computer animators is the selection of a graphics board that complements the capabilities of the host. Generally speaking, the host is best suited to perform such tasks as file transfer and data manipulation. An on-board graphics processor can compute and display complex information directly. In this way the board takes a load off the host and bypasses the bottlenecking of data processing routed via the host. The real-time alteration of part of a computer graphics image is a good example of a situation in which a graphics board's limited intelligence can outperform the host processor.

Trade-offs abound. A problem typical of boards with a lot of on-board intelligence and greater resolution and color capability is that interpretation of the incoming information can take longer than it would simply to display the image. The best measure of a board's real-world speed and overall usefulness is a hands-on test and comparison to other boards using real application software. Some software may take advantage of a dumb board, making for superior performance in terms of speed. Another software package may take advantage of multiprocessing by directing the calculations associated with complex shading (or some other computationally intensive task) to the graphics board.

The safest general conclusion is an almost self-evident one: Boards with a higher CPU clock speed are generally faster than their slower predecessors. However, don't be fooled by speed specs alone.

Video Interfacing

For the animator, one of the most popular features of a graphics board is the ability to interface with the real world via video. If video is your medium of choice you should evaluate your options carefully so that you aren't constantly upgrading. (For a detailed discussion of video, see Chapter 13, which focuses specifically on matters relating to video production.)

Graphics boards with dedicated video output can incorporate various levels of output quality that should be carefully assessed. A board advertised as video compatible, for example, may be able to produce video only to the level seen on an ordinary video TV monitor. That's simply not good enough, however, for many professional applications. To get the quality you need, you may need a converter or other extra features to make the board's output recordable on a professional home VCR or a professional VTR.

Video output from the graphics board to the display unit can be generated in either red, green, blue (RGB) or the composite video format (these topics are covered more extensively in Chapters 6 and 13); many boards support both formats. *Composite video* is a standard format that is used to support ordinary video cameras and VCRs. Composite video utilizes a single coaxial cable, usually with standard RCA connections (like those found on a typical home stereo) to transfer video information. *RGB*, in contrast, requires at least three cables for the three colors and additional cables for horizontal and vertical synchronization. Typically, the five lines are bound together into one thick cable for convenience.

Composite equipment (such as video recorders and monitors) is easy to use, but the level of quality is unsuited to most professional applications. For higher quality output, RGB signals are preferable because they produce sharper images and fewer color artifacts. Another popular format is S-VHS (also known as S-video or super VHS). S-VHS is a more recent standard that interfaces with newer VCRs and video cameras. In terms of quality of output, S-VHS is superior to composite but does not approach the caliber of RGB.

Special Features

If video is to be the final format of your animation, you need to be aware that common synchronization signals must be consistently maintained throughout the system. If the sync is not correctly matched, the board will display video at a rate slightly offset from that of the video input camera, the VTR, or the other video components. Therefore, to ensure sync, you should consider graphics boards that offer genlocking.

A graphics board that incorporates *genlocking* has the ability to generate its video scanning timing from an external sync source. Here's how it works: A VTR (a potential source of sync) sends both vertical and horizontal sync signals to the input of the graphics board. The board, in turn, makes sure that the incoming timing signal coincides with the outgoing signal. In other words, it locks onto an external timing signal and thus produces output that is synchronous with it. Without genlocking capability, a VTR would attempt to record a mistimed video; this would result in an offset and thus unviewable frame. Genlock is essential for single frame video recording.

An additional feature common to superior video-oriented boards is the provision of an *alpha channel*, also referred to as a video overlay plane. This feature is especially useful when you want to add video effects to an animation or to use video in conjunction with 2-D painting or animation. The alpha channel can be used to mix live video with a given image or sequence of images. This means that you can display a computer generated image on the board and at the same time display live video by assigning a particular color—black, for example—to be transparent. In this way, you superimpose live video input on the animated images. The intensity or amplitude of the live video is more or less controllable, depending on the depth of the alpha channel, as specified by its number of bits.

With depth beyond 1 bit (where the video is either all the way on or all the way off), an external video image can be gradually increased or decreased in intensity to produce gradual fades and other effects. A depth of 8 bits, for example, provides 256 (2^8) levels of brightness. The alpha channel facilitates rotoscoping (this is discussed later in Chapter 13) and painting over video. If you want to create a painting that replicates a live video image, you can transmute black into transparent video. That way you can see enough of the video image to paint over it. Turning off the alpha channel (via the paint program) reveals only your painted image. In such a case, the painting can also be fashioned to fit actors' positions. A woman's face, for example, can be displayed so that different hairstyles and hats can be painted over her original videotaped image. Later, after the video is recorded and played back, the painted digital images will be perfectly registered with the woman's face.

Other video capabilities are turning up on the newer graphics boards. Some of these are designed to facilitate manipulation of animated images. These features can expedite video production by easing the addition of special effects and transitions, and the compositing of multiple video sources. One of these features is worthy of special note. *Chroma-keying* allows the animator to relegate any color to transparency in order to display live video from an outside source. Features such as these can significantly boost your video animation capabilities.

Graphics Monitors

6

The monitor is one of the most important components in an animation system simply because it is the most visible part of the system—it formats your view of the work in progress. The monitor (sometimes referred to as a cathode ray tube, or CRT) is of special importance to an animator because it is a tool for development, a previewing medium, and in some cases the final showplace for the creative work. For these reasons, the choice of monitors should not be treated casually by the animator: Their attributes must be well understood.

Unless the monitor is well understood, a catastrophe, rather than a masterpiece, could be in the works. The monitor's functional importance becomes obvious when its technical capabilities are misunderstood or somehow mismatched. For example, if the target animation is to be seen on video but the animation has been created using a large RGB monitor, the differences in format between the two can be considerable. The aspect ratio (horizontal to vertical size) may be incongruous, and colors can be problematic (some are not possible on broadcast TV). Another typical problem is that parts of the RGB image might remain undisplayed when replayed on video.

Because each type of monitor is designed with a different purpose in mind, it is not uncommon to see many monitors in use at a production facility, each dedicated to a specific function. However, this discussion focuses on monitors that are engineered for graphics (not text) display. First, you'll take a brief look at how a monitor works, and then you will use this knowledge to take an informed look at monitor selection.

The Monitor from the Inside Out

The computer delivers to the monitor a series of signals that are translated into dots on the screen. When the graphics board is providing RGB signals, the monitor uses these separate signals (and two more for the horizontal and vertical synchronization) in order to compose the

Adjusting for Disaster. Here's a little cautionary tale that illustrates the importance of the monitor. At one of the better computer animation studios, it happened that a client was viewing a sequence for approval. While the chief animator was elsewhere, the client felt compelled to adjust the composite monitor's contrast and tweak the hue, without the producer's knowledge. Later, after the animation had been completed, the client returned to the editing room for a viewing and noticed, much to his dismay, that the video animation looked remarkably flat and the colors were shifted. This kind of bungling can easily occur in any shared production environment. The solution is to tape over the monitor's setting; this helps to remind everyone that the controls are not to be adjusted. In this case, the result of the client's simple adjustment required a reproduction of the animation.

screen's image. Without sync signals, the RGB monitor's electronics would be unable to coordinate the display of the color information.

The amplitude and timing of the electronics are critical. The amplitude determines how brightly a given color should be displayed, and the timing indicates when the color should be on screen and for how long. The vertical sync signal tells the monitor when it is time to start a full screen cycle. The horizontal sync signals serve to synchronize the display of each horizontal line of graphics information. This horizontal line is called a *scan line*. Together, the five distinct signals describe to the monitor's electronics how and when to display a particular color at a given horizontal and vertical time. As you might expect, denser images are more computationally intensive because of the greater number of dots. Thus, the greater the number of pixels that require definition, the faster the monitor and graphics adapter must work together to display them adequately. Because this task can be taxing on the electronics, one of two alternative methods is generally used to accommodate the display of color information: Monitor displays are either interlaced or noninterlaced.

The *noninterlaced* strategy is the easiest to understand: The horizontal scan lines are displayed sequentially from the top down (see Figure 6–1). Then the vertical sync pulse indicates the start of another

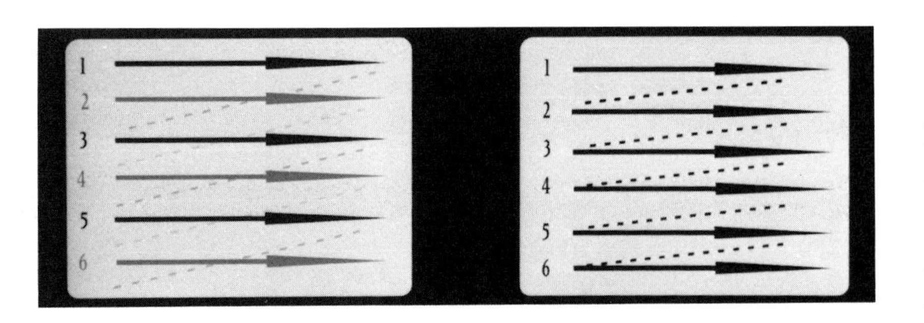

Figure 6–1. *A graphics display unit and the monitor can be either interlaced (left) or noninterlaced (right). Both units, however, must use the same scanning method. The interlaced method shown here uses alternate scanning, and the noninterlaced method uses sequential scanning.*

frame, which typically occurs 60 times a second, or at a rate of 60 hertz. For high resolution, this process requires extremely fast electronics to drive the monitor at a high voltage (more than 24,000 volts for a 19-inch monitor). The result is a quality image; however, it is generally an expensive one to manufacture.

That's why the interlacing technique was developed. In an *interlaced* display, each scan line is alternated so that odd and even lines are displayed serially in rapid succession. In this way, images are displayed in only half the vertical resolution (called fields); this means that the electronics components have to work only half as fast as they do in a noninterlaced monitor. A noninterlaced display displays both fields—a full frame—serially. A good example of an interlaced monitor is a typical home TV set. Each 60th of a second, the viewer sees an even field at half the resolution, followed by the odd field. The switching is mostly imperceptible.

Color monitors use a special (CRT) tube, which is used to display colors in a proper configuration. Inside the tube is a metal grid with holes (called a color mask), which directs electrons toward the phosphor in the correct region for each particular color (see Figure 6–2). The space between the holes is referred to as the *dot pitch*. Smaller dot pitches are preferable because they detract less from the detail of an image. Dot pitches on currently available monitors range from 0.41 millimeters to 0.21 millimeters. The position of the electron beam must be exactly calibrated; otherwise, other colors (in addition to the assigned one) will be illuminated. This quality, referred to by manufacturers as convergence, is a function of the monitor's electronics that guides electrons in their path to the phosphor.

Although all monitors are RGB on the inside, some monitors are specifically designed for specialized uses, as indicated by their respective signal types, such as composite and S-VHS. Composite monitors are designed to utilize a signal that has encoded the separate RGB information together with sync information. As you might expect, the resulting

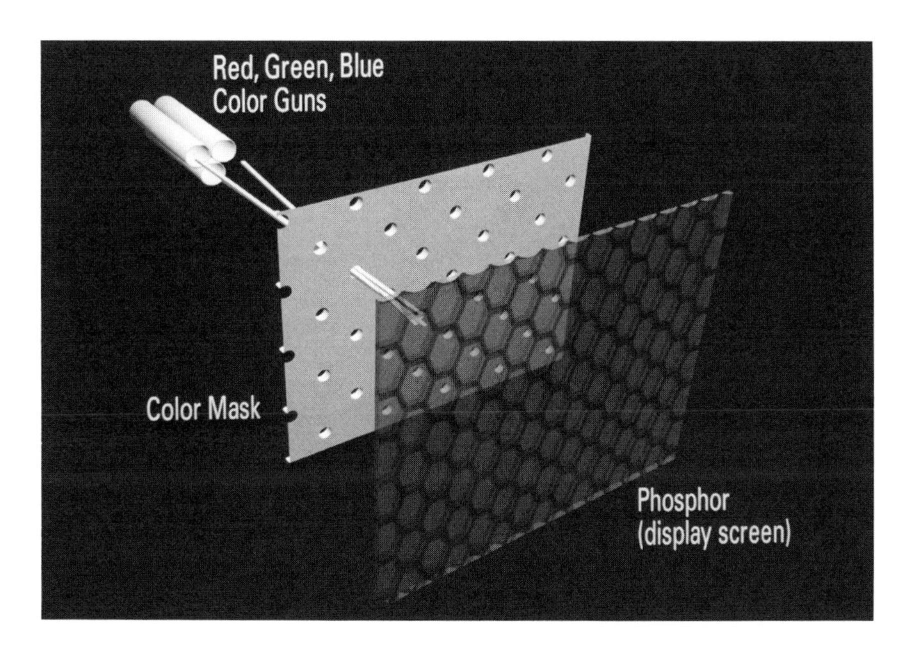

Figure 6–2. *The dot pitch of a color mask is based on the distance between the holes in the internal metal plate. The smaller the distance, the finer the resolution.*

signal and image is inferior to standard RGB and is used only for TV or video applications, never for high-resolution applications. In a similar fashion, S-VHS also needs to be decoded before it reaches the RGB electronics. Still another type of signal interface standard used with computer monitors is called *transistor-transistor-logic* (TTL). TTL signals don't vary from 0 to 1 volt, as do analog input signals. Instead, they are on or off, switching from 0 to 5 volts, which is typical of computer signals. Given the current popularity of computer graphics applications, many monitors on the market incorporate some or all of these standard input types.

The Monitor from the Outside In

Selecting a monitor is a complex and ultimately personal endeavor. Prices, which often (but not always) reflect quality, vary dramatically, and some important features typically go unappreciated by many users. For the animator, the selection of an appropriate monitor is critical. Qualities that need to be evaluated are as follows:

- type of monitor
- resolution(s) and (displayable) size
- interlacing
- dot pitch
- convergence
- visual and geometric linearity
- color purity
- stability
- extra features

These qualities together define the operational characteristics of the monitor.

The monitor you select will, in effect, configure your working environment, and so it's not a matter of simply going out and purchasing the best monitor on the market. As you have already seen, different monitors are optimized for different tasks. The monitor must be selected to match your particular (usually multiple) needs as a computer animator. In fact, a typical animation setup utilizes at least two monitors; one high-quality RGB monitor, which is used to create the animation, and a composite monitor, which is used to display relatively poor workaday video. A caution is appropriate at this point: Be careful when placing one monitor in close proximity to another. Because of the large electrical fields generated by monitors (color monitors in particular), monitors can affect one another considerably, and so they should be placed at least two feet apart.

You need to be able to anticipate the kinds of differences in appearance that are likely when an image is displayed on different kinds of monitors. For example, you will notice that many colors are simply unobtainable on a TV-grade monitor. A gold watch, for example, may look like gold on an RGB monitor but have a hideous brownish cast when viewed on a composite monitor. For this very reason, many animation facilities position two contrasting monitors next to one another so that sequences can be viewed in different formats at the same time.

When working with a composite monitor, you should remember that the viewable area is restricted so that you see less than what you really have. You won't be able to see the complete image available to those viewing the work on a monitor with overscan, as it is sometimes called. In professional applications, *overscan,* and *underscan* can present some serious problems (see Figure 6–3). That's why professional-grade composite monitors are equipped to allow the viewer to switch overscanning on and off. For similar reasons some monitors can be set up so they display the whole image inclusive of the black border.

A monitor's resolution is a potentially confusing specification—a fact exploited, inadvertently or otherwise, by many manufacturers. Suffice it to say that a great deal of misleading information is available to anyone who investigates monitor capabilities in today's marketplace. For example, you might think that a 1024 × 768 resolution 13-inch monitor with a dot pitch of 0.31 mm can display 1024 × 768 pixels. However, when you do the math, you realize that the small size of the screen cannot display a single pixel at that high a resolution.

When evaluating a monitor's resolution capability, you need to consider the other important attributes of the monitor. Is the monitor interlaced, for example? All capabilities must be designed so they work together to match the graphics board's output.

Interlaced and noninterlaced monitors cannot be used interchangeably. Newer monitors may possess the ability to switch between scanning methods via auto-sensing circuitry that determines what kind of signal is being received. However, most monitors are designed to accommodate only one signal type.

In order to maintain competitiveness with new graphics circuit boards that are capable of generating different and higher resolutions, many monitors currently on the market offer automatic resolution-sensing circuitry. This is a very useful feature. But here, too, you must make sure that the board's low resolution capability (as controlled by the software) is matched to the monitor's capability. Another problem associated with monitors with a variable scan rate is that the size and

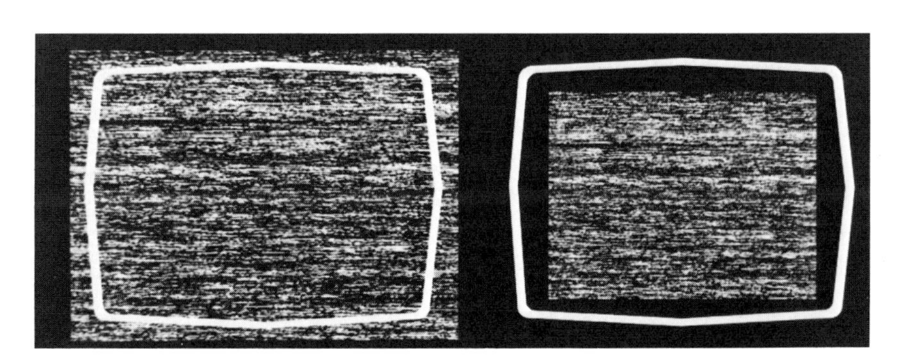

Figure 6–3. *You should be aware of whether the monitor is in underscan mode (right) or overscan mode (left). When a monitor is in underscan mode, you can see the complete image, including its border. When a monitor is in overscan mode, the image will appear the way the viewers will see it.*

aspect ratio (height to width) often change with the resolution. Thus, a circle at one resolution will appear as a small oval at another resolution.

A monitor's dot pitch specification is an important index of display quality. A dot pitch as large as 0.41 is difficult to work with. You will find that images on such a monitor are unclear and poorly detailed. A dot pitch of 0.31 or smaller provides for superior and more useful resolution.

Trinitron-type monitors (pioneered by Sony) are an exception to this general rule. With these kinds of monitors, there is no dot-pitch specification. Trinitron-type monitors use a different method of electron-beam focusing that employs long, thin vertical bars. This design, which enables monitors to provide brighter colors, is preferred by users who work in brightly lit rooms. Some animators avoid Trinitron-type displays because the vertical lines of the monitor can interfere with the vertical lines on the graphics image.

The degree of a monitor's convergence can seriously impact the computer animator because of its effect on detail work. If a monitor has poor convergence, what should appear as a white line may in fact break out into off-set red, green, and blue lines. In CAD applications, in which line's color often has a specific meaning, this can be extremely frustrating. You may notice that, despite seemingly satisfactory convergence specs, a monitor may show symptoms of poor convergence, particularly toward the edges of the display. Another symptom of poor convergence is that colors may be off. Problems such as these should not be tolerated by exacting computer graphics professionals.

It should be apparent by now that specifications are sometimes misleading and so are only useful as rough guidelines. There is no substitute for a careful hands-on and visual inspection of a monitor's attributes.

A similar problem exists with linearity. A monitor's linearity can be expressed as a geometric measure (such as how straight a line is) or as an electronic measure. Variations in the linearity of the electronics can exist. For example, if the three color amplifiers are not properly tuned and synchronized, a full-screen image that is supposed to shift from black to white will appear bluish at one region and reddish at another. Likewise, if the amplifiers are in poor alignment, the black portion may suddenly shift to a medium grey, or perhaps to a bright white. As you might imagine, working with these kinds of linearity defects can make it nearly impossible to create a photorealistic image.

Geometric linearity can be equally frustrating. When square objects appear crooked and circles appear as ovals (see Figure 6–4), drawing is complicated and it is difficult to anticipate the final appearance of an computer generated sequence.

Another issue you should consider when choosing a monitor is color purity. In this respect, you need to first understand that no monitor displays pure colors. Each monitor is unique; a monitor will show a color response different from that of a monitor of the same kind made by the same manufacturer. You—and your clients, if you are doing professional work—must accept this fact. There are, however, some predictable tendencies. Certain types of monitors have color personalities; Trinitrons, for example, tend to be brighter yet display more orangelike reds.

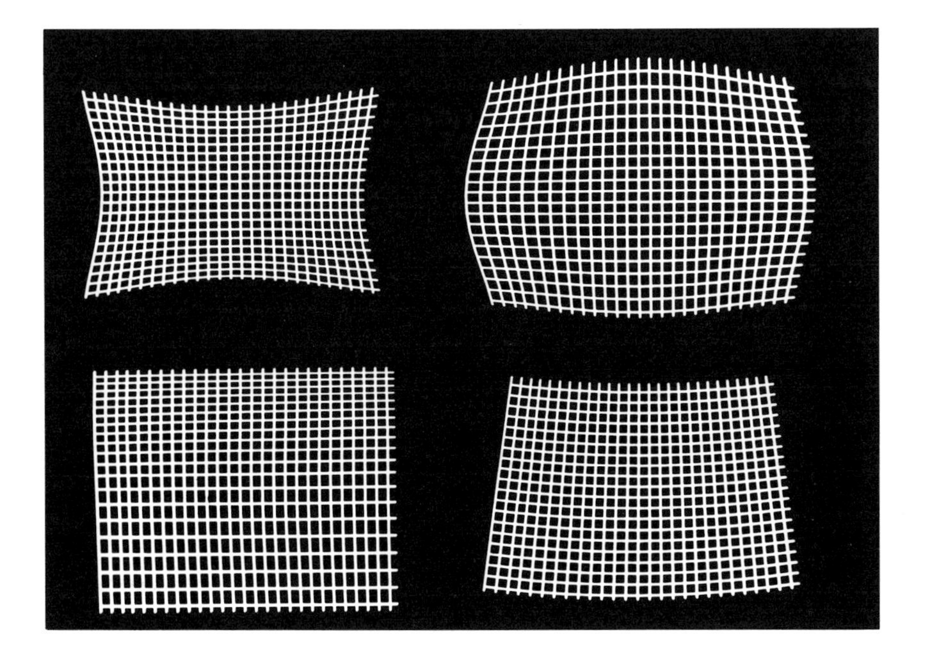

Figure 6–4. *Geometric distortion, such as that shown in these examples, can falsely bias your perception of the image.*

Color quality may not be displayed consistently in all areas of your screen. To test a monitor's color quality, you can create a display of white and then observe exactly how white it is throughout the display. You may discern "warm" spots or differences in color or brightness. Many factors subtly affect a monitor's overall color purity. You'll find that colors are likely to change with temperature and are also affected by the age of the monitor. That's why monitor manufacturers recommend that you wait at least half an hour before using a monitor in a color-critical application—give it time to warm up to a stable temperature.

Power line fluctuations can affect a monitor's brightness, color, and geometry, all of which affect your ability to work with in-progress images. If the quality of your power is a problem, you should consider installing a power regulator with a tight voltage specification (such as under 5% regulation).

Finally, some key features incorporated into many monitors can make your life as a desktop animator a good deal easier. For example, a monitor with adjustable vertical and horizontal image size and position, in addition to the other standard electronic adjustments, allows you more easily to manage the artifacts resulting from changes in resolution. Another feature to look for is loopthrough. The *loopthrough* feature allows you to use other monitors in tandem at the same time. A second set of connectors is provided that allows the video signal to be routed in serial fashion to another monitor or monitors.

Adjustable termination facilitates loopthrough and is a feature that requires some understanding of electronics. *Termination* is an electrical message that signals the end of a chain of devices. It says, in effect, the end is here. If, for example, three monitors are to be daisy-chained together, the last one must be terminated in order for the monitors to work together. Without proper termination, the signal levels of the monitors are incorrect and the on-screen images so generated are functionally unusable. Termination is said to be adjustable when it is switchable—alteration of termination status is usually effected via a switch on the back of the monitor.

Connecting the Monitor

Because of a lack of standards regarding linkage, a number of different methods are typically used to connect monitors to graphics output devices. In come cases, the link is easily accomplished via a single cable. Occasionally you will need a number of different cables. Cabling for synchronization signals can utilize any of the three following methods:

Option 1. The vertical and horizontal synchronization signals can be separate, requiring two additional wires along with the three for red, green, and blue.

Option 2. The sync is (or can be) composite sync, and so the vertical and horizontal signals will typically use only one cable.

Option 3. A single color (usually green) is also used to convey the composite sync information, and so only three wires are used (the other two carry red and blue).

Differences arising out of sync signals can cause a cable nightmare when different monitors, film recorders, and/or graphics boards are being switched around. For this reason alone, it's a good idea to identify their purpose and/or destination by labeling cables with the device (or devices) they connect (colored masking tape works well).

If separate cables are used, cable lengths need to be the same. If they are not, the times the signals take to get to the monitor will be slightly off-set and so will the dots on the screen. As you might expect, this kind of sync problem can adversely affect image quality.

From this review of monitor features it should be clear that monitor selection is an important, sometimes difficult, and yet essential task in the assembly of a functional desktop animation system. As a final note it is worth repeating that specifications are useful only as guidelines: The only dependable way to evaluate a monitor is to test it.

2-D Animation Techniques

7

Earlier chapters scrutinized the hardware tools required to create computer animations; this chapter focuses on the software that mediates your interactions with the equipment. Currently available software products that facilitate desktop computer animation ride a constantly moving wave of improvement, upgrading, and technological enhancement. Many of the capabilities to be found in animation software tools have evolved out of the experience and need of professional production companies and are ultimately based on current animation trends. New techniques are spawned daily through a kind of technological synergy that blends conventional techniques with flourishing hardware capabilities. This palette of ever-expanding capacity provides the nascent animator with a huge toolchest—mostly untested—that awaits your exploration and use.

Computer animation production can be grouped roughly into two clusters, 2-D and 3-D. Although the underlying rules and conceptual points of departure may be shared by these methods, the techniques, viewing audience, technology, methodology, and finished look are decidedly different. Some animators and software manufacturers view 2-D as a subset of 3-D capability. Others see 2-D as an excellent animation production process in itself, as new and unique capabilities evolve to support the creative requirements of the advanced animator. Sophisticated computer animators are knowledgeable about both methods and often derive techniques and tools from both camps in their day-to-day work. This chapter focuses specifically on high-quality 2-D animation software and the techniques associated with its use.

2-D animation can be thought of as the sleeping giant of computer animation. Compared to 3-D as a working mode, it is considerably less constraining, and it is cheaper, faster, and allows more creative license. So why isn't it more popular? First, 2-D is often thought to be limited to cartoon-style animation because it works with flat objects having only height and width. Second, the process of creating 2-D animation is based more on artistry than on technology. Many visual attributes, such as shadows, are automatic with 3-D animation, but 2-D animators must use their artistic and creative skills to create the best look. For this reason alone, many animators quickly overlook 2-D in favor of the automation that comes with the use of 3-D techniques.

2-D animation techniques can provide unique solutions to creative problems. 2-D provides unlimited possibilities for visual compositing or combining graphics from different media. Imagine a background composed of wildly moving colors. Candy cane palm trees appear out of nowhere and land on a beach made from live video that was processed to look like it was hand-drawn. Photographs of real clouds float in the sky, and hand-drawn characters move about in the foreground. In the hands of a creative 2-D animator, such images can be powerful, arresting, and effective.

Here's a brief comparison of 2-D and 3-D animation:

2-D Animation	**3-D Animation**
Limitless visual capabilities	Constrained to a 3-D world and program limits
Manual shading, lighting, and perspective	Automatic shading, lighting, and perfect perspective
Manual description of movement	Near-automatic motion control
Mixing of various techniques	Cautious blending with limits
Selective visual treatment of graphics elements	Can't treat objects differently
Fast development of preview and animation	Slow process of creating models, previewing, and animation
Special fast tricks	All tricks and effects are slow
Can mimic complex 3-D scenes within limits	No limit to the camera's position
No limit to scene complexity	Realistically limited to some level of scene complexity

Due to the unique power inherent in both 2-D and 3-D animation, their capabilities are often imaginatively combined to produce effective animated sequences. Animators will often make a 3-D background (sometimes using it only as a guide) and then superimpose 2-D characters on it. In fact, experienced animators often prefer to work in 2-D, if only for the cost savings. Consider this scenario, for example: An animated special effect needs to be integrated into a video and will last for only about one second, which is only 30 frames of video. In such a case each frame could be created entirely, and very effectively, in 2-D. If you don't really need to use 3-D, 2-D animation is faster to produce and thus less expensive to create.

Examples of 2-D Animation

Perhaps the most prevalent example of 2-D animation is seen daily as titles scroll across the television screen. In fact, broadcast television relies significantly on quickly generated effects, which are readily implemented using dedicated 2-D animation workstations. Imagine this: A gold medallion quickly swings into place (with shadows) while text (letters with their respective shadows) is positioned over a video of a moving pool of water. All of this can be accomplished quickly in 2-D by an experienced animator (see Color Plate 2 for an example of combining several 2-D elements).

2-D animation is also used extensively in TV commercials. You probably would never guess that much of the hand-drawn animation for commercials was electronically drawn and animated on a computer. Parts of a scene, called visual elements, can be moved about, creating a 3-D-like effect. Trees in the foreground, when made slightly larger and moved outward toward the sides of the image, give the impression of a camera going into a scene. Although this may seem cartoonlike, it is a good example of how the computer can automate what was once a slow manual process.

2-D animation is gaining popularity in the corporate world. Business presentations can be dramatically enlivened by use of simple animation techniques. For example, adding movement to bar charts and graphs adds interest and life to what otherwise might be a boring presentation. These kinds of presentations can be created and displayed on most available desktop PCs inclusive of sound. Creating an animated presentation is a straightforward task for the animator because prefabricated images are often available and can be easily set up for animation. Indeed, software programs specially designed for business presentations readily combine text from documents with images (in any number of native graphics formats) and can incorporate sound effects to boot. With the right graphics hardware, the animated presentation can be recorded directly to a common VCR.

2-D Methods

2-D animation is entirely dependent on the software that's in use. You will discover that most available 2-D software products focus on particular aspects of 2-D animation, such as cartoon character development, titling, or creating special effects. Generally, however, the main 2-D techniques include:

- animating paint system functions
- manipulating 2-D objects
- compositing multiple images
- video effects
- rotoscoping

Animating with Paint Systems. The most basic kind of 2-D animation software simply extends the power of a paint system so that painted actions can be recorded and played back. In fact, all 2-D animation systems incorporate a paint system of some kind. A simple paint system provides basic drawing and coloring functions and provides for creation of boxes, circles, lines, and other shapes. Better paint systems provide airbrushlike effects, textures, tinting, image cutting and pasting, multiple fonts, and special effects such as smearing, blending, area averaging, image twisting, image rotating, and image enhancement.

A 2-D animation system provides the animator with the tools necessary to perform all of these kinds of drawing and painterly effects and then provides a means for automating control. These systems have a recording capability by means of which your actions on the paint system are recorded for playback later (at various speeds). This, incidentally, is how animated titles are made to appear as though someone were writing them by hand on the video screen. What's more, a good 2-D animation system will also provide previewing and editing of the timing of the effects you create.

Aside from some workaday titles and business presentations, what's 2-D really good for? In short, a lot. Imagine, for example, that you want to create an animated title with a little flair. Your 2-D animation system can provide the letters, which in this example will be brought in from the side, out of focus. The out-of-focus effect is made by blurring the first frames of the letters. Now you want to dress up the effect a little. You can use streaks of airbrushed multicolored fireworks—they'll all of a sudden manifest, then appear to explode. Getting the same kind of effect in 3-D, by the way, is difficult; out-of-focus effects are generally not available. Now imagine the animated letters glowing brightly and then transforming into a new set of letters as they center, come into focus, and become more readable. All of this can be animated on a painted background. Finally, the last set of letters can expand while the whole scene rotates, turns into a flying sheet of paper, and shrinks to nothing. Without the constraints of a 3-D "reality" there are no rules and no limits except those built into the 2-D animation system, its paint capabilities, and your own imagination and skill.

A feature incorporated into many of the dedicated 2-D animation programs is a kind of rough-and-ready preview capability. This feature enables you to play back your in-process animation directly via the computer's display screen (at various playback rates). What's more, an animated sequence can often be run on another computer via a player program. The *player* is a segment of the full 2-D animation product that is in effect a dedicated playback mode for the program. In fact, manufacturers of these products provide player programs free on electronic bulletin boards as a purchase incentive.

Most paint programs today provide image importing and exporting facilities that support various popular formats. Some popular file formats are PICT and PIC (for the Macintosh), and Targa, PCX, TIFF, and GIF for MS-DOS computers. If you want to transfer images from one desktop platform to another (for example, from a MS-DOS computer to a Macintosh), and your computer is unable to read the disks containing the images, one solution is to transfer the data from one computer to the other via modem over telephone lines. An electronic mailbox, a service available via electronic bulletin board systems, simplifies the transfer process by allowing for temporary storage between uploading and downloading.

Paint by Frames. As stand-alone products, paint systems are often used for animation despite the fact that they lack animation capability. Animators can use them to create a sequence of images, drawing individual frames as needed. This may seem like a lot of work—and it is. However, in some situations it is advantageous to use a fast and capable paint system to create a short animated sequence.

The creation of a lightning effect is a good example. Animated lightning starts with a frame of small, sharply defined off-white sparks. The next frame follows with full-blown sparks airbrushed with glows of color, say reds, yellows, and an electric blue. If there is a background, such as a logo, it is typically tinted to a near-white, indicating extreme brightness. In the third frame, the sparks will have disappeared and the logo will appear nearly normal. Blue glows might still remain, as the lightning is now almost gone. This entire effect can be drawn using only three images, and then it can be manually added onto the video using a high-quality VCR that has single frame editing capability.

Another example illustrates the fluent capability inherent in 2-D animation. Suppose you need complex motion, but you can see that creating it would be too time-consuming. You know that the projected animation will last only a second (30 frames or so), and you're really in a time crunch. The solution? First of all, you can "shoot on twos," thus creating only 15 images and recording 2 frames at a time for a total playback time of one second of animation. In the process, you make sure the effects you use fuse flawlessly from frame to frame. If, for instance you needed to make a cloudscape via airbrushing, you might have difficulties shifting from one frame to the next, and without careful blending the animation might appear mottled. With the addition of sparkling water, however, this effect just might be acceptable.

Manipulating 2-D Objects. A number of 2-D animation software products provide not only the ability to create 2-D objects but also the very useful ability to manipulate them. A 2-D object may be a letter of text or a cartoonlike character, such as a bug. A bug, for example, can be made larger or smaller by scaling it. Such an object can also be squashed and/or stretched—two very important qualities derived from classical animation. This in effect rescales the bug in the horizontal (X) or vertical (Y) direction. Objects (the bug in this case) can also be rotated, moved, or any combination of these effects to convey the sense of action.

One advanced 2-D manipulation capability is the ability to paste objects on an imaginary pane of glass and move it. Scaling the appearance of an object (such as a bug) is equivalent to increasing the width of one side of the imaginary glass pane. Tilting the glass so that the top is farther way from you than the bottom makes for distortion and mimics a similar kind of 3-D effect—it invokes a sense of perspective (see Figure 7–1).

Wilder, more eccentric motions can be simulated when the imaginary glass pane rotates and twists about, producing optical distortions similar to special effects seen on TV. These effects are sometimes called Ampex digital optic (ADO) effects. Additional functions may include controlling an object's degree of transparency, the overlay of other images, and duplication of objects. One advanced feature is the ability to transfigure one object into another. A simple example is changing a box into a star. This process is called *metamorphosis*.

As you might expect, 2-D programs based on object manipulation and painting usually provide a means to control and edit motion. They are typically set up so that the number of frames (of which there are 30 per second in video) can be determined while the animator is manipulating the relevant object(s). A time editor also helps to control the quality of the motion. If you prefer motion that starts slowly and then speeds up—called an ease-in—you simply adjust a graph that displays motion speed vs. time. This feature is especially useful when you want to give lifelike motions to living characters (see Figure 7–2).

Imagine a walking character, for example. For realistic locomotion, the up and down motions of a character's walk should evidence some variation in speed of motion. The walk radiates a character's personality; this is just as true for animated characters as it is in the real world. Up and down motions can be adjusted so that one foot quickly elevates the character, then he pauses at the height of the cycle, and finally he quickly drops to ground level. This automatic averaging between motions or static positions is called *interpolation*. You can either instruct the computer to interpolate speed automatically from one position to the

Figure 7–1. *A single 2-D image can be distorted (in 2-D) and used individually or, as shown here, as a group. The same technique can be used for creating a flurry of rotating, scaled, and twisting snowflakes for a snowy background.*

Figure 7–2. *An essential feature in a 2-D animation system is its ability to control and edit different visual elements in a sequence. (Director Software version 2.0 reproduced with permission of MacroMind.)*

next, or you can intervene and edit the motion to approximate even more lifelike timing.

Most 2-D animation programs rely on a concept developed decades ago in the beginnings of animation. In order to simplify the animation process, chief animators first created *key frames* of the critical positions of a character. Then an assistant animator made all the in-between frames, a process called *tweening*. 2-D (and 3-D) programs operate in a similar fashion. The animator sets up a key frame for an object, say a ball. The ball is first placed at the top of its fall. That's the first key frame. The next key frame shows the ball's position on the ground. The computer then automatically tweens all the frames between the first key frame and the last, according to how many frames are desired. Finally, with a motion editor, as described above, the motion can be made to go slow at first and then speed up—just as it does in everyday life (see Figure 7–3).

The various 2-D animation products have evolved different strategies to control 2-D editing. However, the resulting capabilities are similar to that of a good word processor: The animator can cut and paste portions of the movement of an object or objects so that they can economically be modified and/or replicated. Using a timing display that graphs time vs. objects (commonly referred to as a channel list or motion editor), you can see what object is in motion at any given time. Indeed, a good graphically oriented time editor is an invaluable tool.

Imagine taking a time-line graph of a particular object, say a jumping frog, and then duplicating it to create the display of an additional frog. Naturally, doing this several times more creates a lively hoard, without your having to do any more work. Making an additional frog, painting it black, and duplicating it in the same way provides a way to generate instant fake shadows. Of course, the frog's shadow must be made to appear "behind" the colored frogs by overlaying the foreground frog images over the shadow images. It's also a good idea to spread them about. This is done by clicking on each on-screen frog and placing it where it belongs. Now that you have a group of frogs, with their

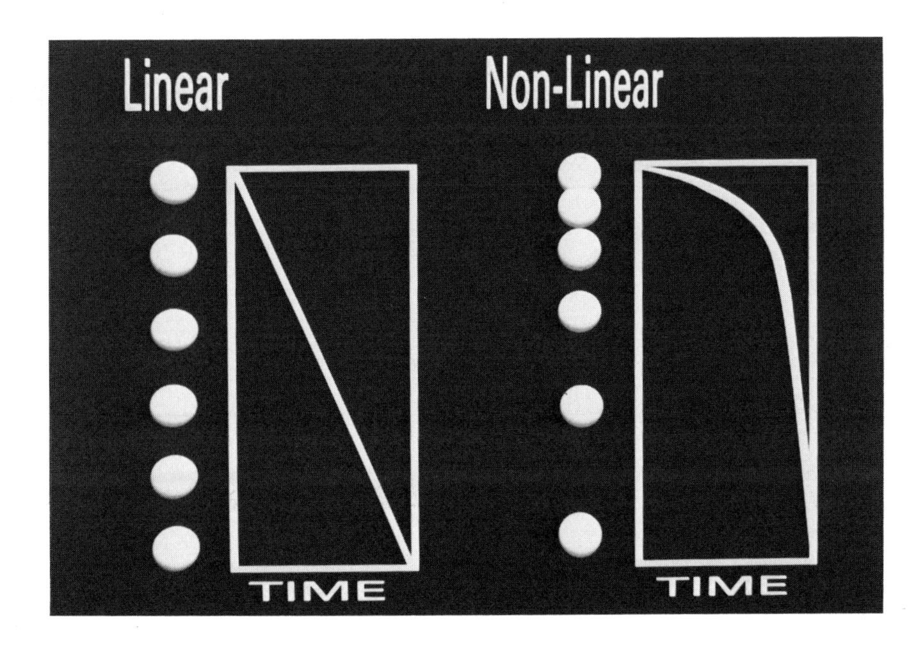

Figure 7–3. *Control of the acceleration of objects is often provided so that more natural nonlinear motion can be attained.*

respective shadows, if you make the leader jump, the rest will jump too. If you don't like them to jump at the same time, you can offset the horizontal position of the time displayed on the time-line description of the motion editor to stagger the jumps.

Perhaps the most compelling feature offered by dedicated 2-D animation programs, however, is that they are designed to be viewed in real time; that is, the animation can be played back as fast as you want it to be. However, due to the complexity of the manipulation, the software often creates the frames in memory. Depending on the power of the system and the complexity of the animated sequence, this can take several minutes. The software then plays them back in real time. The ability to create an animation quickly and view it immediately is a potent feature. It provides the animator with a way to refine movement by providing immediate feedback before the sequence is finished.

Compositing Multiple Images. A fundamental feature associated with 2-D animation is the ability to combine images. This capability, called *compositing,* is similar to the process of cutting out and merging a series of magazine pictures. However, there is a significant difference: The animator can make the composited image move. Different types of images can be so joined. For example, a composited 2-D animation might be constructed of captured video images, photographs, 3-D synthetic images, 2-D paintings, and any other source of 2-D imagery. What makes this particular process so effective is that the software smooths the edges and thus serves to fuse diverse images into a completely new one. For example, a black object placed on a white background produces sharp, jagged artifacts at the edges. The smoothing process, called anti-aliasing, averages these edges into gradations so that image quality is greatly improved.

Another useful feature included with some 2-D animation packages is resolution-independence. This capability allows you to select the final resolution for the images and animations you create. Thus, with this feature your 2-D animation is not necessarily limited to the resolution you see on the display screen. This allows you to create in low resolution and evaluate the quality of your work as you go (which you can do by playing it back in real time). Later, you can rerun the animation at a much higher resolution, or at its optimal level of quality. Indeed, some programs are capable of running out a sequence at a resolution suitable for film (which can be as much as 4000 lines of resolution). Of course, as you might expect, when you are producing at this level of resolution you won't be able to view even the simplest animation in real time.

Video Effects. Many 2-D software packages currently on the market address the needs of video production, especially production of titles as well as other basic video effects. Although limited in scope, these kinds of specialized 2-D animation products can be an important and useful addition to a complete animation system. A typical product focuses on creating moving text. Usually, a wide range of text fonts are available, and these fonts can be colored, scaled, moved, scrolled, and imbued with transparency. In fact, you'll find in such a package most of the features found in a commercial video text generator. Many of these products include essential drawing capabilities as well as basic video features, such as fades (making a transition to a color, typically black). Some allow the animator to select areas transparent to video and provide accurate timing of activities as well. Because all these graphics functions operate in real time, these products can be important allies to the animator who is intent on producing professional-looking video.

Some dedicated 2-D animation products target traditional animators, and so the manufacturers fashion their software in ways that are familiar and sensible to the classical animator. For example, using this software approach, a character outline is drawn first and saved. Then another drawing is created and superimposed on top of the previous one. The software visually suggests the progression of the assembly by subtly shading the most recent drawing so that its outline is slightly darker than the previous one. In this way the animator can see the earlier character outlines as fainter and fainter images underneath the current drawing (see Figure 7–4). This ongoing sequence of character outlines provides an excellent overview of an object's overall movement.

Rotoscoping. Rotoscoping is commonly used to get the feeling of live 3-D movement in 2-D. *Rotoscoping* is the process of obtaining moving information from a so-called real source, such as video or a motion picture of a particular action. For example, the animator might wish to use a video of a person running as the basis for a character's movement in a certain sequence. Using the video source, the animator can trace the action frame by frame—in effect outlining a sequence of movements that will become the basis for an animation. Although there are many ways to rotoscope, the following example uses a graphics board with a video overlay. These provide the animator with the ability to project video onto the screen and simultaneously paint over it. Here's how it's done.

The software enabled the animator to draw objects and details while at the same time viewing the video as a background and guide. The animator isolates, or freeze-frames, an image on a VTR so that it can be

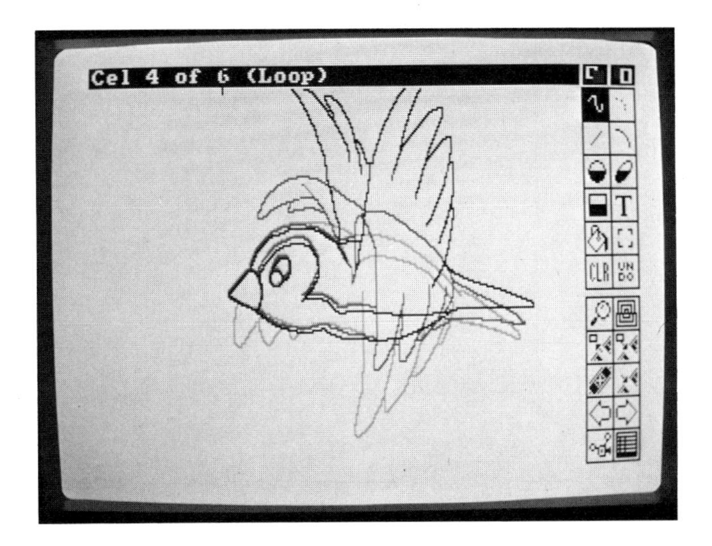

Figure 7–4. *A useful feature based on classical animation techniques is called an onionskin effect. It allows you to view the last few frames simultaneously so that you can draw the next frame accurately, thus enabling you to create smooth motion. (Reproduced with permission of Disney Software's The Animation Studio [bird], and Gold Disk's Animation Works [man].)*

viewed on the graphics board as live video. Then, using a paint system, the animator outlines an object's movements. Once this outline is completed and saved, the next frame from the VTR is called up and the process is repeated. After a sequence of frames has been saved, the animation software can be made to play back the images in real time so that the animator can see the outlines created (without the video) by tracing over the single video frames.

Examples of 2-D Animation Tricks

The best examples of animation in 2-D are produced on an animation/paint system. Furthermore, they attest to the fact that good animation is based on highly creative use of graphics tools. One trick exploits the large virtual drawing area of the paint and 2-D animation MS-DOS-based program called *Lumena* (from Time Arts). The paint program allows you to create an image larger than the computer screen can display. In fact, you can put several of these images side by side to make a massive background. Why would you need such a large background? If you want to have an animated character running in front of a background that doesn't repeat itself (until a certain point), you can have the software pan across the background, and you can place the sequence of the running 2-D character in the foreground.

Mac Animation Example. As an example of a 2-D animation on the Macintosh, consider *Director* from Macromind. The software is based on the creation or incorporation of Mac-based images and sequencing them together. The images can be small or full sized. A small image can be scaled, rotated, or twisted and then moved about. The background of the small image is transparent so that it doesn't obscure the actual background of the complete scene. Many of these small images can have visual priority over others so that one may overlap another in a predetermined way. What makes the product unique, however, is that the animation sequence can be graphically edited. A chart of the "characters" and their activities are displayed and then edited so that the timing of any of their individual activities can be changed, moved, elongated or replicated. This can also be done with sound effects which are a part of the software. A product such as this is excellent for live animations to be viewed on the Macintosh such as software demos, corporate demonstrations, etc..

MS-DOS Based Animation. An example of an IBM-PC based 2-D animation product is Animator, from Autodesk. The graphics are intended to be displayed on any of the VGA graphics boards, but those which have video output capability allow the animation to be recorded on standard home VCRs. The software includes a paint system so that images can be drawn and then manipulated to create a wide range of effects. The software provides importing of a large number of image files so that 3-D creations or video-captured images made on a Targa-16, for example, can be used as a part of the sequence. The images are stored on the hard disk and played back as fast as the disk will output data, which is often faster than video's 30 frames-per-second rate. A system such as this provides rapid animation development capability which can then be used for video presentations.

Building a 3-D Model

8

The 3-D world is magical. The most exciting developments in computer animation have been in the realm of 3-D—a fact that's not all that surprising when you think about it. Much of the work associated with 3-D animation is both repetitious and predictable, and so the field is a prime candidate for what computers do best. Indeed, computers have made dramatic inroads into 3-D drudgery. The resulting animations can be made to appear stunningly photo-realistic, and, when coupled with surpassingly smooth camera motions available via software, the output can eclipse even the most advanced products of traditional cinematic animation.

In fact computerized 3-D camera movement allows the animator access to locations heretofore impossible to reach. Once the door to the 3-D digital world has been opened, to display a given cast of objects, you (the virtual camera-person) can go anywhere within the confines of this world. You can make any object in it move in any way you choose. Artistically, you are virtually unlimited—once you've created the 3-D environment.

The art of 2-D animation is based on compositing as well as on deliberate and detailed articulation of character movement. As you will see, 3-D animation techniques provide depth and considerable automation to the animator's toolkit. When you move into the 3-D domain, you'll find you have access to a new and powerful dimension. You'll also find that there is a price to pay for the extra dimension: 3-D imposes new disciplines and very exacting ones at that.

Synthesizing models is the most obvious drawback to working in the realm of 3-D animation. You will discover that creating the digital models of objects, whether they be words or cities, can consume as much as 90% of your time. Its tedious manual labor: The more detail you incorporate in a model, the more points you must add and the longer you and your computer will take to define and construct your model. Remember, all the detailing of an object must be absolutely correct. That's why it so important that you become really familiar with your 3-D modeling program: You need to know how to access all its capabilities well enough that you are free to spend your time efficiently on 3-D model construction.

Like the professional filmmakers, who have developed a thorough working understanding of camera technology and lighting, you, the computer animator, need to understand fully the nuances of your software and to have a comprehensive understanding of the full capabilities of your hardware. As you construct your 3-D world, decisions about lighting, control of movement, object interactions, and the positions assigned an imaginary camera will all seriously impact your finished product, initially in unpredictable ways.

Although 3-D animation programs differ substantially (even from version to version) with regard to model making and the animation process, the fundamental procedures that produce a 3-D animation remain basically the same. Here, step by step, is a basic description of how a 3-D animation is put together:

Step 1. An unambiguous 3-D digital model is created in a 3-D space. The 3-D objects as a group are called *models,* and the software that enables you to do this is called a *modeler.*

Step 2. The model is verified for accuracy by rendering it as solid objects. Note that this process requires full descriptions of surface characteristics of the model, camera position, and environmental lighting.

Step 3. 2-D backgrounds are created with a paint program (or can be made in 3-D and converted into a 2-D image).

Step 4. The model (each object with its respective background image) and the virtual camera's point of view are given a path of motion over a specified time frame.

Step 5. The motions of all the objects are previewed—in low resolution or in wire-frame—and edited as needed (see Figure 8–1a).

Step 6. The finished animation is then saved in memory or rendered frame by frame onto video or film (see Figure 8–1b).

This chapter focuses on step 1, the animation aspects of 3-D, that is, on modeling and not necessarily on 3-D still-image generation. Other books and journals deal admirably with the intricacies and ever-changing

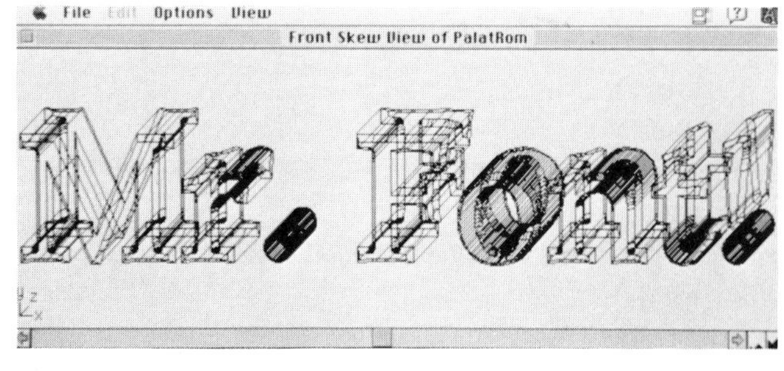

a

b

Figure 8–1. *To create a 3-D model text, (a) first choose a font (based on 2-D shapes) and then (b) extrude them into thick letters. (Electric Image Animation System software reproduced with permission of Electric Image.)*

details of 3-D still-image generation. Succeeding chapters focus on steps 2 through 6. You will see how 3-D objects are rendered with 2-D images, how the camera and lights are included, and how the animation is scripted for the movement of the objects and the camera. Finally, in still later chapters, you will see how an animation is imaged onto video or film.

3-D modeling methods vary depending on the purpose of the application. Some applications, for example, require slicing through an object to reveal its interior qualities. This method is called *solid modeling*. In most 3-D animations, for example, for video and presentation purposes, only the exterior appearance is required. This book focuses primarily on *surface modeling* methods.

Because of its complexity, 3-D animation software varies in its overall functionality. Some programs provide only modeling or rendering capabilities. Others provide a complete 3-D modeling, rendering, and animation system but do not provide painting capability. As a result, many animators use several 2-D and 3-D products to meet their particular preferences and requirements. Because of the wide variation of features offered in currently marketed animation software, you should closely examine available products to make sure that they adequately address your needs.

3-D Space

3-D models exist in a world possessing height, width, and depth. These dimensions are placed on a grid called a *Cartesian coordinate system*. The horizontal line is the *x-axis*, the vertical line is the *y-axis*, and the line indicating depth is the *z-axis*. The center, or *orgin* of the three-dimensional system is described as the triplet (0, 0, 0), which denotes the locations of x,y, and z, respectively. Anything to the right of the x-axis is considered positive, anything to the left of it is negative. In a similar fashion, the positive values for the y-axis are upward, and the positive values for the z-axis approach you, the viewer. For computer graphics purposes, the numbers used are arbitrary and can be thought of as feet, millimeters or any other unit of measure. The use of a mathematical system such as this means that the objects and their component parts are always in an explicitly defined location based on three assigned numbers.

Building 3-D Objects. A *3-D object* is the product of a set of descriptive points that establish its position and conformation in a Cartesian coordinate system. A modeling program displays such an object in wire-frame representation with lines drawn to connect its set of points.

For example, six triplets (x, y, z) adequately and completely describe a cube's shape and position relative to the confines of the coordinate system.

As you might imagine, creating a set of points for every object in even a simple animation is incredibly tedious and time consuming (yet, in the past, that's how it was done!). Modern 3-D modeling programs have evolved an number of alternative methods. Perhaps the easiest system to manage is one that starts with so-called primitives. In this system, the animator utilizes already-created primitives, which are graphic models of basic shapes. In a 2-D scheme, these objects are, for example, simple circles, squares, and rectangles. In a 3-D world, however, the primitives are cubes, cones, spheres, and the like. Once you have selected a primitive such as a sphere, you can stretch it, squash it, and blend it into other, original shapes. You might even choose to combine several primitives to create objects with more complex shapes.

Another way to create a 3-D object is to start with a 2-D object, which already possesses height and width. The outline of a 2-D object can be described as a set of points that are joined together. A square, for example, is a set of four points with a single line connecting them. This closed line is called a *polygon*; the polygon is one of the essential components of 3-D computer graphics. The polygonal square, in this case, can be given depth, a z-axis, by the process of extrusion. When a polygon is extruded its set of polygonal points is duplicated and offset and then the two sets of points are connected to create, in this example, a cubelike shape (see Figure 8–2). You'll find that most modeling programs allow you to extrude a 2-D object from various angles—you aren't limited to a vantage straight out from the face (which, by the way, is called the normal).

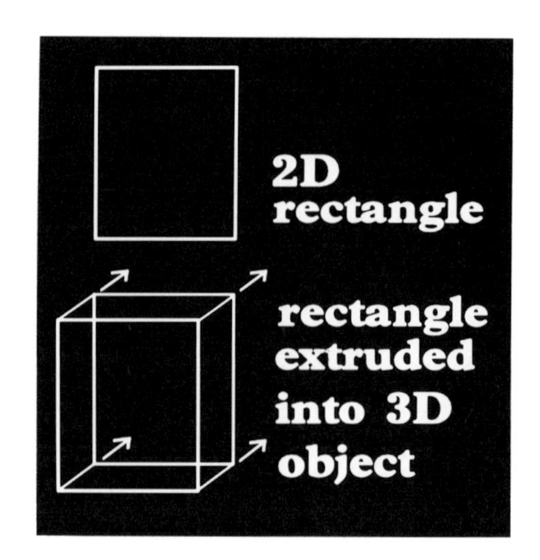

Figure 8–2. *Extrusion is the process of taking a 2-D shape and giving it depth. The modeling program often provides a choice of depth, angle, and curved extrusion paths.*

Now look at a more practical example of extrusion. Say you wanted to turn a lettered logo into a 3-D object, extruding it to the depth desired (just as you would with any polygon) (see Figure 8–3). You would first load in a selected style of polygonal letters from a modeling program (a range of fonts are usually provided). Next, you would need to assign the set, or angle, of the letters to be extruded. Many programs let you look at the letters from various vantage points, from the top down. Thus you can establish a position for the depth of the letters, and you can select the angle if you want the extrusion to be offset.

Say you also want to show only an outside band surrounding the letter. The editing function in the modeler lets you choose any particular polygon and remove it. In the case of Figure 8–3 the original faces of the letters are removed. So, you remove the original fronts and backs of the letters and are left with only the exterior, extruded part of the letters. You can create dramatic effects such as this using simple extrusion and editing techniques.

Figure 8–3. *A more sophisticated effect can be given to simple blocky letters through beveling.*

During the modeling process, editing activities allow you to change a model considerably. You can edit points, 2-D shapes, or 3-D objects. Simple editing functions include *translation* (moving), *scaling* (in any direction, *x*, *y*, or *z*), and *rotation* (from any central axis). You can usually add, delete, or move individual points or lines as well. A more advanced feature offered by some programs is called beveling. *Beveling* is a process similar to extrusion whereby you create one face rather than two for any given angle. The slanted edge can be made either rounded or flat. *Beveling* is typically used to enchance simple text, giving it an elegant appearance. The letters possess angles that accommodate complex highlights and reflections when animated.

Other advanced features in a good modeler facilitate the editing process. Sometimes models become so complex that you can't adequately identify and specify certain sections because too many polygons obscure your target object. In such a case you can *hide* an object, making it appear or disappear as needed. Hiding one or more concealing objects allows you to see and work more easily with a target object. If you want to know the locations of the hidden objects (to use them as reference points), you can specify them to display as bounded boxes. When you do this, the objects are represented by simple exterior lines, such as that of a rectangular box, instead of as hundreds of polygons. This feature allows you to move objects about with relative ease because there are fewer lines to display and thus fewer for the computer to calculate.

A rounded object such as a vase or bottle is created by first constructing a polygon resembling half an outline of the envisioned object. You then perform what's called a *surface of revolution* around the half-polygon's axis. In Figure 8–4, a wine bottle is modeled by generating half of the side view and then rotating the outline on the vertical axis along one edge.

Figure 8–4. *Using the surface of rotation function, a shape may be rotated around an axis into a 3-D shape. Here, a half outline of a wine bottle is transformed into a 3-D bottle.*

3-D models are composed of facets. A good example of a many-faceted sphere is the mirrored ball seen hanging in old ballrooms. Each mirror in the mosaic is a facet. In a 3-D model (of a sphere or any other object), the grouping of polygons that make up an object is called a *polygonal mesh.* The combination of numerous flat of polygons can make an object appear relatively smooth—indeed, the denser the mesh (the greater the number of facets), the smoother will be its appearance. Later you'll see that the surface properties of an object can also affect the appearance of the facets.

The number of facets (flat faces or polygons) assigned an object that are used to produce the desired effect is an artistic decision. The optimum number is whatever you think looks best. For the wine bottle of Figure 8–4 you could choose a surface of revolution utilizing only three sides or one composed of fifty sides. As you might expect, however, there are trade-offs to model complexity. The greater the degree of complexity, the longer is the animation time. That's why it makes sense to work with simpler models having fewer facets.

A set of logical operators allows you to add or subtract from previously produced objects in order to generate more complex objects. Using the *logical operators* AND, OR, and NOT AND (referred to as Boolean operators), you can discriminate in the selection of objects or parts of objects. In Figure 8–5, a sphere is cut via subtraction by rectangular blocks. The advantage of using logical operators in your modeling is that you can subtract part of an object—say a cylinder residing inside a cube—and then add parts from different models (products of other logical operations) to create entirely novel single objects of great complexity.

Figure 8–5. *3-D objects can be subtracted from others: This sphere was subtracted by a group of rectangular blocks.*

You can accomplish this process in either 2-D or 3-D. You can create the teeth on a gear, for example, by first creating a circle and then creating a square. The square can act as a cutting tool, subtracting its polygonal overlapping from the circle. When this has been done, you can extrude the flat gear into a solid-looking object. The same object could also have been made by creating a circle, extruding it, and using a cube to cut away at the solid circle.

Rotating Confusion. When you transform an object around itself, as in a rotation, you need to realize that the software assumes that the axis of roation is the object's center. If this is not the case, the center should be assigned a different axis. Also, multiple rotations on different axes of an object can be confusing. As you rotate an object, take careful note of the steps you take so that you can backtrack if necessary.

Another method can be used to create a complex curved shape from an assembly of polygons. Suppose you wished to create a model of the hull of a boat. First, you create the ribs of a boat: Using a modeler, you create a series of curved polygons, where each rib is slightly different depending on its position in the hull. Then, when all ribs are in place, you connect the polygonal ribs into a single solid object, a boat hull. This process, called lofting, is typically used to fashion very complex shapes.

Creating Better Curved Surfaces

You will find that there are times when the use of polygons does not result in a satisfactory solution to model building—the model may need too many polygons to make this a practical approach. This brings us to another method that can be used to create a 3-D model, one which is often better suited to constructing and showing complex curved objects.

This method is dependent on the use of curved lines—called splines—which are generated by mathematical equations that define how a curved line should be drawn. These are not the point-to-point descriptions of lines but rather formulas based on polynomial equations ($y(x) = ax^3 + bx^2 + cx + d$). (Don't worry, you don't have to do any of this math; that's the computer's job.) Different types of splines can be used: Some intersect the dots that determine the course of a curve, and others create the best curve while never touching the dots. It is not necessary to know the mathematics behind these curved lines, but it is helpful to know the advantages and limitations of each type. Figure 8–6 illustrates how a rounded square is made where the control dots are integral to the polygon.

Cubic equations are used to form what are called parametric surfaces, or bicubic patches. Control points on the splines can be used to define the shape of the curved line of the patch. Control points are like handles on the surface or object; you can twist or pull on the handles, and the effect is similar to stretching a rubber patch into new shapes (see Figure 8–7). Making an object this way allows you to create complex shapes—such as a twisted airplane wing. Moving the control points can change the fineness of the tip, the curve of the trailing edge and the twist from the thick part to the wing tip.

As a model maker, you can use either of these two primary modeling techniques—polygonal and patch-based—so you will need to

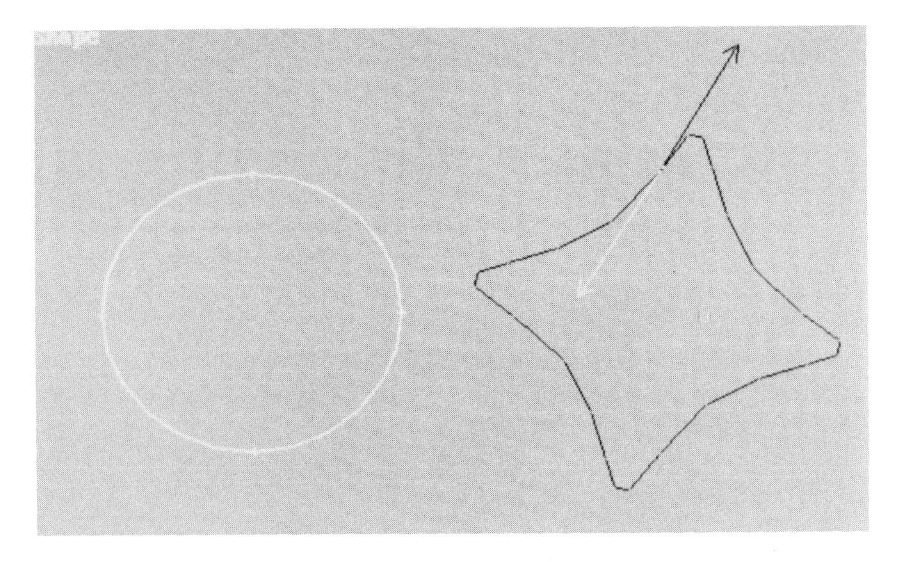

Figure 8–6. Spline-based circles can be dramatically transformed into different shapes via their control points. Changing the direction of the arrow, for example, can turn this into a circle.

determine which method is suitable for the creation of a given model. If both methods are available with your modeling program, you have the additional option of combining methods as you describe an object. In general, if the object is of complex shape and rounded, such as a car hood or a human face, the patch-based method will produce superior results (see Figure 8–8). Objects produced with the patch-based method are often easier to edit into the best shape. Polygons work well, however, for simple geometric objects such as walls, logos, and such.

Grouping or Linking Objects Together

While creating objects of increasing complexity, you may find that grouping them into sets or organizational groups is helpful. For example, if you want to duplicate the attributes of a single house to make several

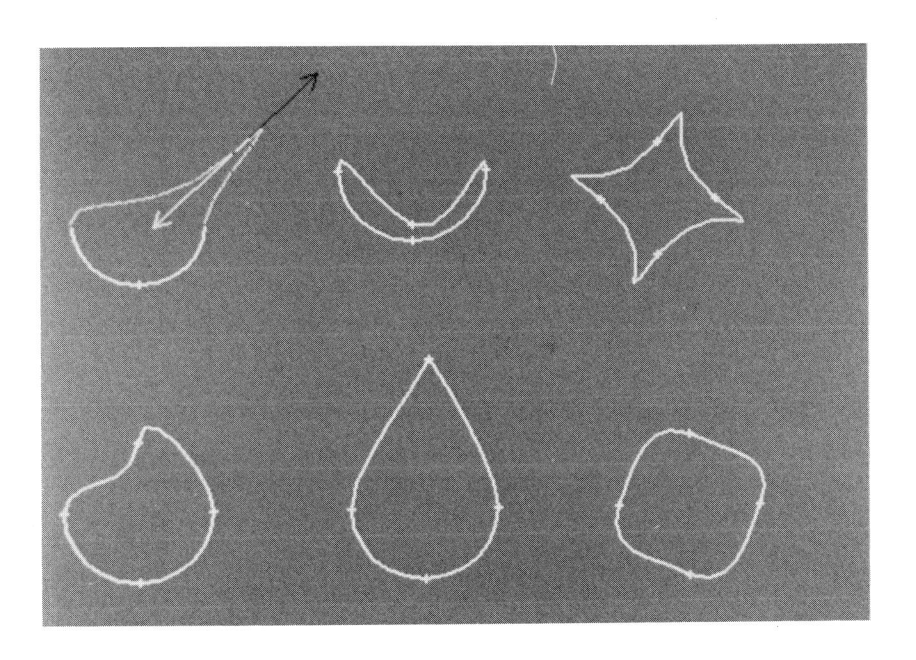

Figure 8–7. Using the bend and twist functions, 3-D objects can be made into completely new objects.

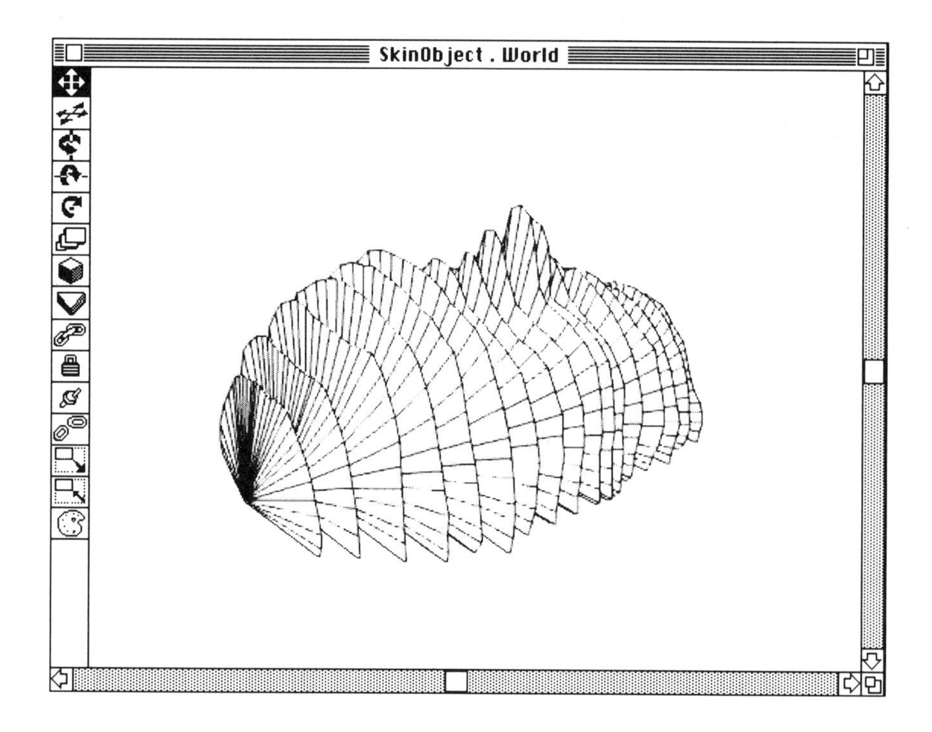

Figure 8–8. *Individual polygons can be arranged as a (contour) group for creating complex surfaces such as a face. (Reproduced with permission of MacroMind.)*

houses, you'll find that if you assign the parts of the house to the whole house, you can facilitate later movements. You can do the same to the furniture. If you want to replicate furniture in one house for inclusion in another (assuming the same floor plan), you can designate the furniture of the first house as a group and later duplicate it and move it to other houses, saving an enormous amount of work.

The same is true when a model is to be animated. Some objects will have relationships with others. For instance, a bat and ball may be grouped together during some parts of an animation but not during others. How such objects interrelate needs to be planned in advance. Groups of objects can be set up so that when one object moves, its whole group moves. This kind of planning can be critical to the success of an animation.

If for example, a set of plates is to move at a particular time in relation to another object, you may find it difficult to organize the movement of the plates during the motion scripting process.

The process of grouping relies on what is called a master-replicant or master-slave relationship (also called parent-child). One object is chosen as the master, and replicants (duplicated versions of the master object) are then attached to it. The animation or editing process can be set up so that when the master is moved, all of its replicants move along with it. Now if a group within this group—a subgroup in effect—needs to be moved, one of the replicants of the main group must be assigned as a master of the subgroup set of replicants.

In this way, a displayable (on screen) hierarchy is developed. Complex hierarchical models can often be printed out so that the details and relationships of the model are clear. As an aid to this process, some programs provide a method of clarifying the hierarchical relationship.

When the stylus is near an object that is a part of a group, the text screen highlights the name of the object (you can assign or change the name) and its connection to the master group. Regardless of the method used, the specification of workable hierarchy groupings is an important modeling step, particularly with complex models, and must be clarified prior to the creation of any complex object.

Creating 3-D Models via Alternative Sources

An exercise at this point will help to illustrate the various facets of a typical modeling project. Let's say you wish to create a 3-D logo. The logo is to be based on the formation of increasingly complex polygons—the extrusion and editing of which is likely to become an arduous and lengthy process. Because of this, we want to look at shortcuts and time-saving alternatives. One such technique is video digitizing.

Many modeling software programs make provisions for the video capture of 2-D objects (such as a company's logo) and can automatically extrapolate a polygon based on the digitized information. The software is designed to sense the embedded data by tracking contrast differences relative to the background. Because such programs are designed to approximate the most likely line on the basis of the video image, you'll find that retouching is often required. At any rate, to continue with the logo, after you have retouched the polygon or polygons into a satisfactory 2-D line drawing, you can then extrude it into a 3-D object.

CAD systems are a source of highly detailed and complex models. Originally developed to produce blueprints of objects or buildings, CAD database programs that define these kinds of models can be converted from their native format by many computer animation systems into models functionally similar to those native to the animation program. The two CAD formats in most common use on personal computers are DXF and IGES. Use of either of these may, however, require some editing, coloring, or other modifications, but they can serve as excellent raw material because of their accuracy. Incidentally, the reverse is possible as well. Animation programs often can be exported from native format into popular CAD formats, thus providing a means to share models between different programs and even between different desktop platforms (for example, a transfer from a Macintosh to an MS-DOS machine).

Another potential model source is electronic bulletin boards. Accessible by modem, bulletin boards often store and distribute public domain databases of countless objects (planes, faces, cathedrals, and the like). These can be downloaded (often at no cost) and then converted by anyone with a modem and a little time. The data are encoded in any one of a number of different 3-D formats, including Super3D, Mac3D, Swivel 3D, FACT, Pro3D, Wavefront, 3DGF, and Sculpt. However, regardless of format, you need to be sure that the image you select is indeed free to use. Some of these image collections are distributed as shareware and so may require a fee for use.

Model Viewing Methods

Objects fashioned for 3-D use are initially created in wire-frame representation because when only the lines of an object are displayed, computations for scaling and so forth are more easily managed by the computer. There are various ways to view your wire-frame model, and this facilitates the model-making process. The problem is that you need to be able to determine what is in front and what is in back of the model and at the same time retain a sense of the model's relative sizes. If you want to add a window to a model of a house, for example, you need to make sure it is the correct size while also making sure that you are putting it where you want it—on the front of the house and not the back, for example.

For this reason, most model-making programs provide a multiple-view mode so that several views are simultaneously available. Typically, the screen will be divided into four views: top, side, bottom, and a wire-frame representation of an imaginary camera view (see Figure 8–9).

You can choose to view the image in *perspective* (also called an axonometric view) or not. A perspective view is usually the best view to choose because without it there may be ambiguity concerning what's in front and what's behind. In this configuration, parts of the model that are far away are depicted as smaller than parts that are closer to the camera's viewpoint. This represents the view as we would see it in real life and as a camera would see it. Conversely, without perspective, the parts of the image close up and far away are the same size. As you might imagine, the problem with this view, if it is seen exactly from top, bottom, or a side, is that you loose a sense of depth. What's more, lines are likely to obscure one another as they overlap, making editing of model parts confusing and difficult.

An alternate viewpoint is called an orthographic view. The *orthographic view* displays only the face of the object when it matches the

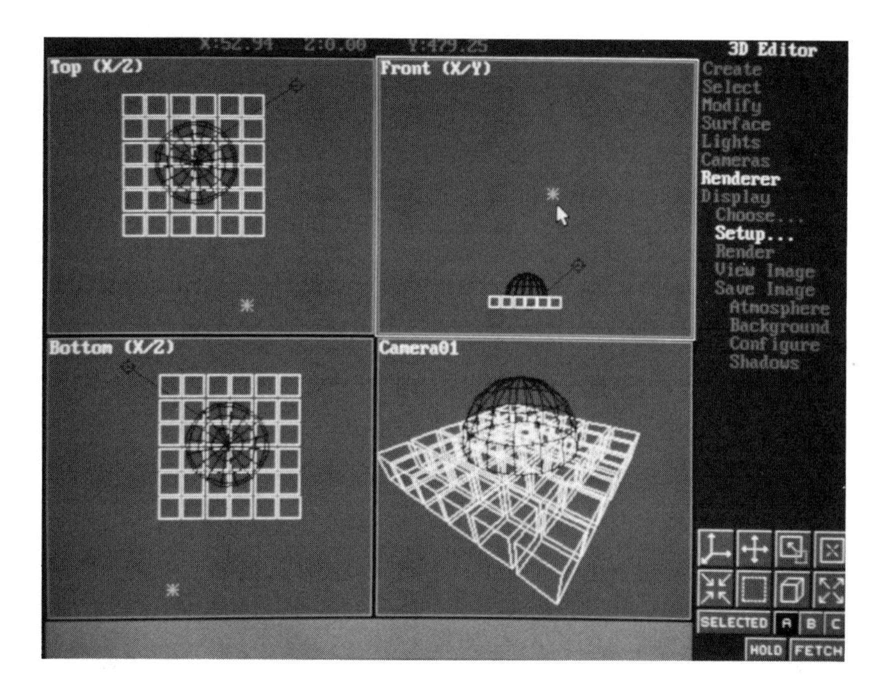

Figure 8–9. *Multiple orthographic views of 3-D scenes offer insight about the construction of the objects. The perspective view, which is how a camera would view the scene, gives you a better idea of how the image will look when finished. (Studio 3D software reproduced with permission of Studio 3D.)*

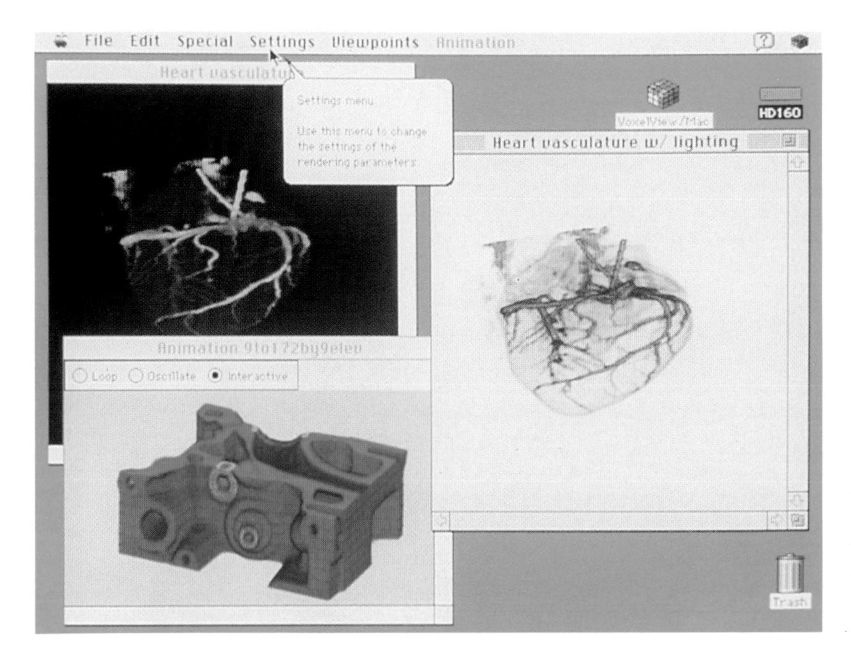

Color Plate 1 *For scientific application, computer animation facilitates the understanding of complex data. (Screen image of VoxelView/ Mac.™ Software by Vital Images, Inc., Fairfield, Iowa. Reproduced with permission of Vital Images, Inc., Fairfield, Iowa; Marcus, Knosp, Frank, Weiss; and The University of Iowa.)*

(a)

(b)

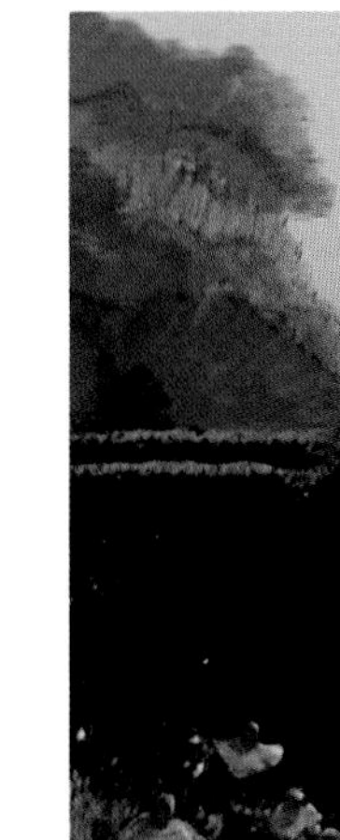

(c)

Color Plate 2 *Several 2-D images can be used together to create a 3-D–like animation. The last image is a composite of several layers of all 2-D elements (such as Figure 2a and b). (Reproduced with permission of Getris.)*

Color Plate 3 *The addition of shadows and reflections adds photo-realism to an image. (Symbolics software, reproduced with permission of Symbolics.)*

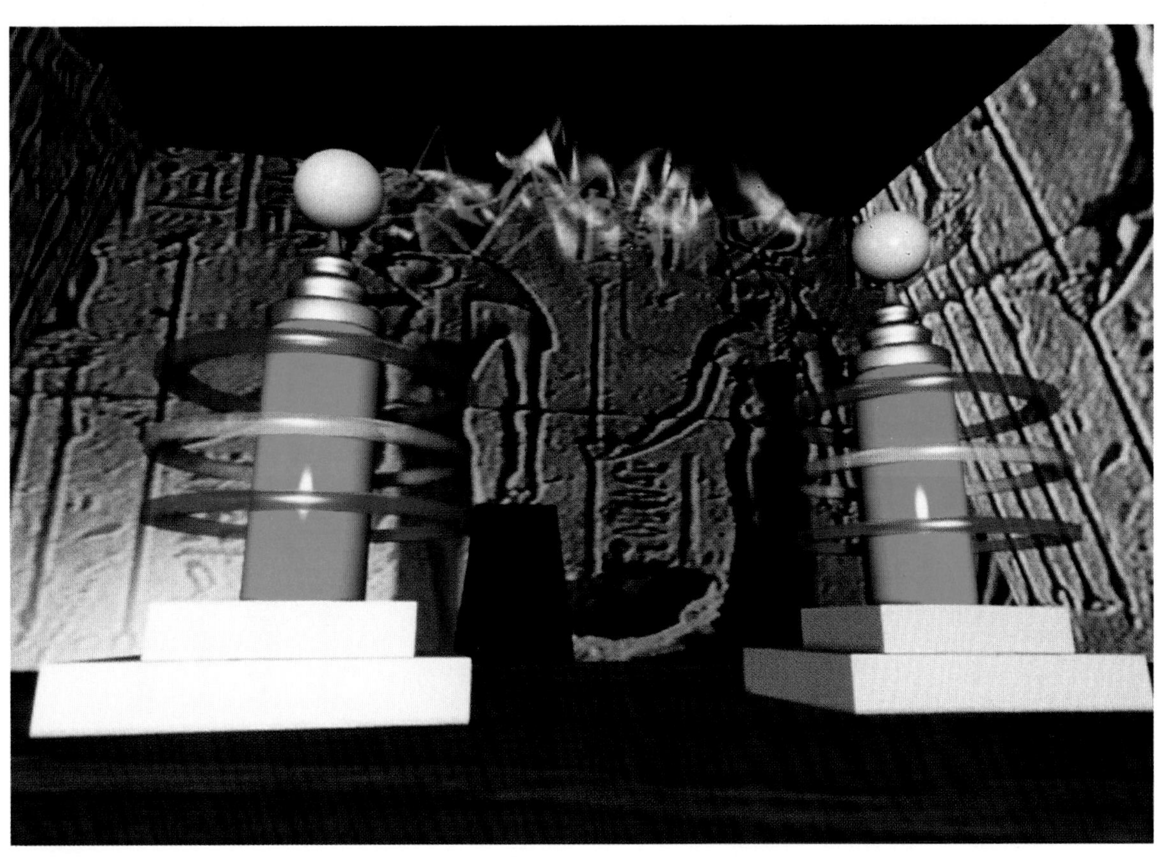

Color Plate 4 *Realistic-looking images can be created by using simple geometric shapes, good lighting, and texture mapping. (TOPAS software, reproduced with permission of Gregory MacNicol.)*

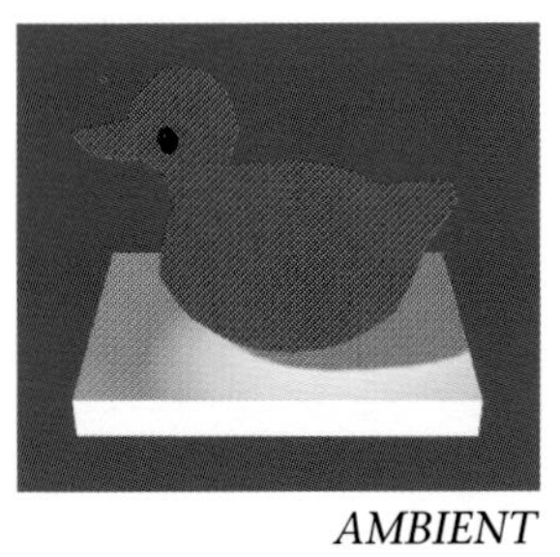

AMBIENT

DIFFUSE

SPECULAR

SHINY9

TILE

DECAL

NEON

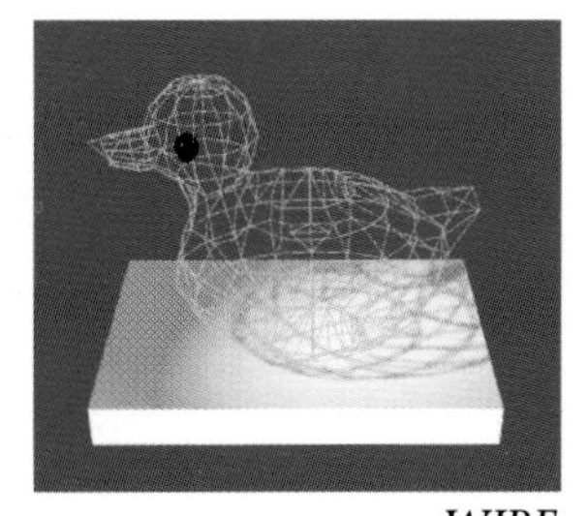

WIRE

TEXTURE

REFLECTION

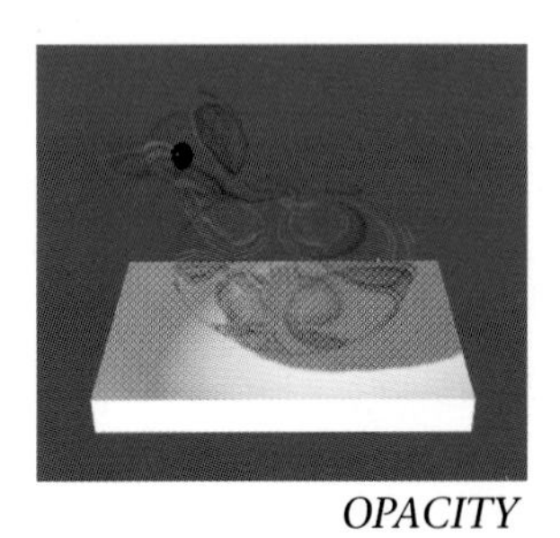

OPACITY

BUMP

Color Plate 5 Texture maps offer an enormous range of creative opportunities. (Reproduced with permission of Autodesk.)

The Materials Editor Samples

SHINY77

ADD

SUB

FLAT

GOURAUD

PHONG

TEXTURE/BUMP

TEXTURE2

TEXTURE/REFLECTION

(Continued)

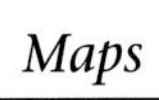

Maps

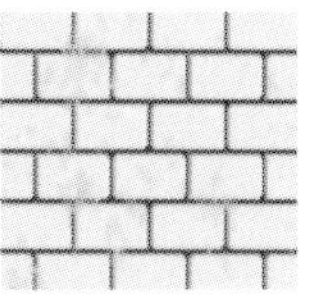

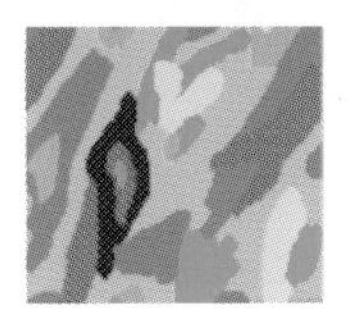

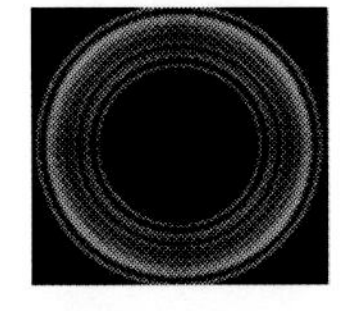

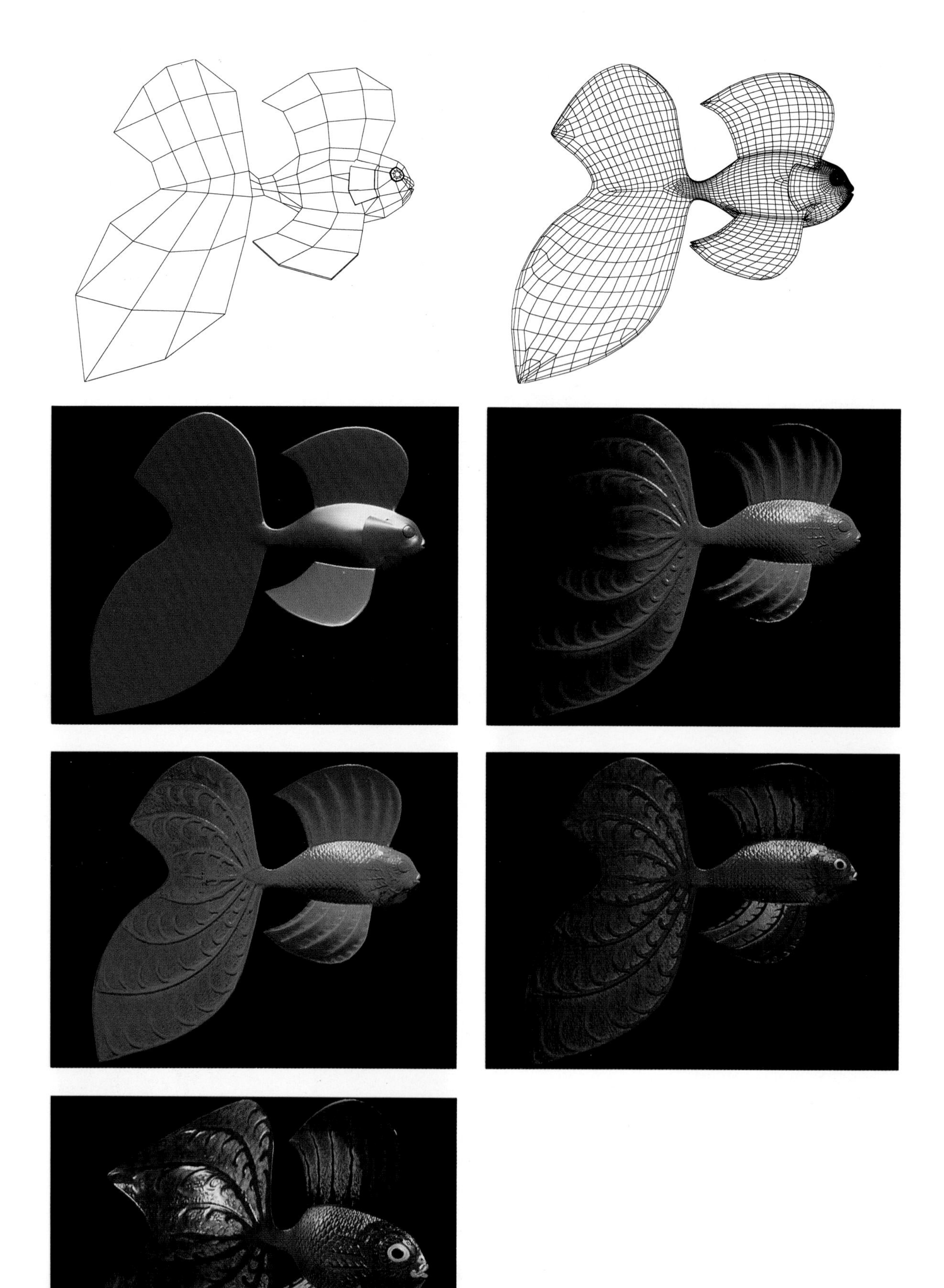

Color Plate 6 *The power of combining different texture and mapping techniques is illustrated by this fish, created by computer graphics artist Matt Elson. Transparency, opacity, texture, bump, and reflection maps were used. (Symbolics software, reproduced with permission of Symbolics and Matt Elson.)*

Color Plate 7 *With good lighting, a completed rendered image can possess a remarkably photo-realistic quality. (3D Studio software, reproduced with permission of Gene Bodio.)*

Color Plate 8 *Most programs offer control over shadows and ambient, diffuse, and specular lights. (DGS software, reproduced with permission of Robert Beech.)*

(a)

(b)

Color Plate 9 *Ray tracing produces remarkable images, despite the fact that the rendering time is exceptionally long. (Reproduced with permission of (a) Ray Dream and (b) Bill Graham.)*

(a)

(b)

Color Plate 10 *The quality of (a) RGB and (b) NTSC is strikingly different, particularly when two colors are placed close together, causing chroma crawl. You should anticipate this when broadcast video is your target medium.*

point of view so it lacks perspective which may appear visually correct but too detailed for understanding the object's orientation. This can be particularly handy for editing complex models such as buildings. This viewing method displays all three (top, side, and bottom) views, which together provide a suitable method for initial model building and editing.

All nonperspective viewing methods are inherently problematic because depth assessment becomes difficult. One feature many software programs now offer to manage this problem is called depth cuing. *Depth cuing* uses the relative brightness of the display—as seen on the wire-frame lines—to signal approximate distance from a given viewpoint. Thus, in a depth cued image, the lines closer to the viewer appear brighter than do more distant lines. In this way, you can tell which lines are supposed to be in front. However, at times even this method can introduce ambiguous elements, particularly when there are many objects on the screen at the same time.

Multiple points of view are critical to effective modeling. That's why most modeling programs provide windows or quadrants on the screen. With these you can display the model with and without perspective using different display techniques. Experienced animators skilled in modeling learn to switch back and forth as they work to obtain the views most advantageous to each stage of the modeling process.

Adding Realism to 3-D Models

9

After your 3-D models have been created, the next step is to give them lifelike attributes. This is an especially exciting point in the process because the addition of realistic attributes allows you to see much more clearly how a scene will appear. You'll find that your earlier wire-frame 3-D representations appear structural and lifeless in contrast.

The process of translating a 3-D model into a more realistic appearing image is called *rendering*. Rendering is a critical step in the animation process. During this process the digital 3-D wire-frame objects (in the form of data files) that were created from the modeling program are made to appear solid and realistic.

Building 3-D objects (as described in Chapter 8) and then assigning them specific attributes is not necessarily a two-step process. Many animation programs integrate these steps so that upon creation of an object, you immediately assign properties to it as you gradually build a complete model. This is convenient during the creation of complex models because you can assign specific attributes to separate polygons of an object. When building an object representing a rooftop, for example, you can at the same time make color assignments. For example, the top polygon can be designated terracotta, the sides dull stucco, and the ceiling (the polygon on the bottom) antique white. This kind of integration is useful because coloring the individual polygons of an object might be difficult later, when your scene includes other polygons that obscure your view or access.

Assigning attributes to objects and their polygonal surfaces is a technical task as well as an artistic one. As you become more familiar with color, you'll learn the best way to blend the colors of an object in a scene. You'll also learn the best way to present the objects efficiently and advantageously in terms of the overall scene. The addition of attributes such as transparency, reflection, textures, highlights, and shadows can turn even the dullest set of cubes and spheres into objects of surpassing photo-realistic beauty (see Color Plate 3). For example, 2-D images of hieroglyphics can be projected onto the surfaces of the inside of a pyramid, transforming the structure's dull, flat interior walls into what appears to be the inside of an Egyptian tomb, as Color Plate 4 shows. This kind of creative effect can be achieved easily once you know how to use your software to assign specific attributes to the surfaces of objects.

For starters, model surfaces (either the individual polygons or a whole object) are typically assigned default attributes. If a given object is a subordinate of a group, the whole group might be assigned an initial color and quality. Imagine trying to decide what color and texture to furnish a real object such as a car, for example. The body can be a combination of dull black, reflective chrome, shiny aqua, and semitransparent. A model of a car, can be imbued with all these conventional attributes; but, if you so choose, it can be made to appear as if it were made of bricks. The digital system on your desktop provides you with a wide range of choices and with the magical ability to make instant changes in attributes.

The resolution of current display systems is limited. As a result, diagonal lines appear jagged, and less so with increasing resolution. To make the rendered image smooth, *anti-aliasing* is performed by the rendering program. Anti-aliasing is a digital filtering technique that is calculated before the image is displayed. Less proficient programs merely smooth or blow the image which is unacceptable. For high-quality anti-aliasing, options in the rendering program provide choices of anti-aliasing values. The trade-off, however, in higher quality is slower rendering speed (see Figure 9–1, a and b).

Surface Qualities

Not every system provides all the possible options that can describe the surface characteristics of an object. Here are some of the primary qualitics most systems will allow you to control:

- color
- surface type
- surface highlights
- transparency
- texture mapping
- bump mapping
- overall quality control

Color. Color is the most basic attribute you can furnish an object's surface. Yet despite its seemingly obvious nature, color is a complex topic. Our eyes register attributes in ways that profoundly contrast with print technologies and computer display systems. Printers use a completely different color methodology than do computer graphics systems. Regardless of your skill and experience, when it comes to color, there is much to learn. Animators can benefit from some of the better technical and artistic books that specialize in color theory and practice. Our

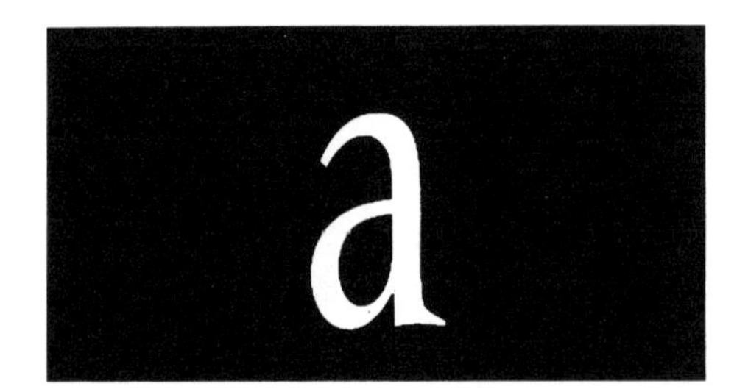

a

b

Figure 9–1. *(a) Anti-aliasing improves the quality of the diagonal lines and edges in a computer generated image via mathematical filtering. (b) Without anti-aliasing, animated objects appear to have crawling artifacts on their edges. Good anti-aliasing is essential for quality animation.*

coverage of color here is only cursory and is intended only as a preliminary guide for the beginning 3-D animator.

The assignment of a particular color to the surface of a model is an aesthetic issue as well as a technical one. To select a particular color from all those available requires some knowledge of the color methodology (called color models or systems) you normally use.

Internally, computers operate in a color mode based on three primary colors: red, green, and blue. Mixing these together produces the range of colors available on your system. It is important, however, to understand that a given color model is a unique perception. It is not intuitive: Not everyone thinks of yellow as consisting of equal amounts of red and green. What makes blue? What colors should be mixed together and in what proportions to make brown? There are no right answers to these questions.

However, artists and others who are not computer users have developed systems based on the quantification of color attributes. One such color system is called *HLS* (hue, lightness, and saturation). The model is based on the process artists use when mixing colors to create new ones. The hue, the basic color, is chosen from the spectrum, or rainbow. Its perceived brightness—lightness—is based on how much white is added to it. A color is also described by the degree of density or saturation. A similar model is called HSV (hue, saturation, and value).

Because these systems are inherently imperfect solutions to color description, most programs use more than one method to assign color. Ultimately, your selection of a method will be based on your experience, your preferred method, and your program's ease of use in the realm of color.

Color systems are useful as long as you recognize their drawbacks. For example, neither color display systems nor print systems have the ability to display all available colors. What's more, color systems are unable to account for human perceptual differences. Therefore, even though we can fabricate machines that incorporate convenient formulas and electronic controls, our eyes are distinctly nonlinear in operation. Psychologists have demonstrated this with any number of perceptual tests: It turns out that each of us perceives differently the colors we designate as red, green, and blue.

An excellent and realistic example of the problematic nature of color description comes from the realm of video. Video formats in both the United States and Europe impose consequential limits on color availability. Because the video signal is encoded for broadcast transmission, the signals are bound by the limits of the medium. In 1953, the United States adopted the National Television Standards Commission (NTSC) system for transmitting color. Called the YIQ color system, it was created to accommodate bandwidth limitations of electronic equipment. The Y designates luminance or brightness, and the I and Q represent parts of the color signal that transmit signal phases. You don't need to understand the technology behind these decisions, but you should be aware that the resultant standard severely limits the color spectrum when compared to the range and subtlety of color available on the RGB color monitors typically used in computer graphics applications.

What this means to you as an animator is that when you are

attempting to create appropriate colors for video purposes, you should view your production using an NTSC monitor, one with color capability similar to what you would see from an ordinary television or VCR. When translated into standard video, highly saturated colors typically appear relatively colorless and bright. Likewise, you'll discover that some colors, such as green and red, possess unacceptable artifacts when placed adjacent to one another on video. You'll also find that some colors, such as gold and brown, are difficult to produce. These tendencies are due in part to biases in the YIQ color model, which leans in the direction of flesh tones—as it was intended. Before the standard was adopted, video tended to give people a yellow or greenish cast.

Figure 9–2. *The three basic types of surfaces provide choices in rendering time and overall quality.*

Surface Type. Surface classifications are commonly used to describe how an object will reflect light—an attribute that can have a significant effect on an object's appearance. A model's shading characteristics may be assigned as flat (also called constant), Gouraud, or Phong (see Figure 9–2).

Flat shaded objects look as you would expect them to, diffused and consistently lighted over their surfaces. Generally, you assign flat surfaces on objects that are small or that don't require variation of brightness over the surface. Text is often "flatted" so that it stands out (for better legibility) instead of appearing "real." Flat shading speeds production because it is the fastest to render—fewer computations are required for the individual pixels because each one possesses the same brightness level. Flat surface characteristics are generally assigned when an image is previewed (because of speed), for making fake shadows, or when few colors are available via the graphics display board (as is the case with an 8 bits-per-pixel board).

Flat shading can have its creative uses as well as its practical ones. For example, if you want to retain the highest rendering speed yet require as a background a large field consisting of dozens of stars, you can form a number of minute flat-shaded rectangles, which when viewed on video will appear as small illuminated objects—that is, they will look like stars.

A faceted object, such as a sphere made of polygons, will display all its facets when flat shading is used. Thus it would have an appearance similar to that of a geodesic dome, which is not likely to be the intended effect. However, by using a different shading model, you can cause the sphere to appear completely smooth and round.

Gouraud (named for its originator and pronounced "goo-ROW") shading is a technique that produces "smooth" shading. It can dramatically enhance an object's surface characteristics and allows interpolation (a form of averaging) between the polygons or facets of rounded objects. The trade-off here is that the use of Gouraud shading imposes a time penalty: It takes longer to render than flat shading. For the most part, Gouraud shading is a very effective technique; however, in certain lighting conditions, objects may have an unwanted plastic or dull appearance.

Phong shading (also named for its inventor) provides even better control, and so, as you might expect, using this technique imposes an even greater time penalty in terms of rendering time (sometimes as much as five times). Phong shaded objects reflect the light more realistically, because they highlight the source of the light. Phong shading is superior because it computes the value of light for diffuse (scattered) reflection and specular (direct) reflection for each pixel instead of for each polygon. It is especially useful in the description of curved objects. Gouraud shading has a tendency to produce the telltale facets of the separate polygons that compose an object (often revealing sharp angles where one polygon was joined to another). An animator might thus choose to use Phong shading to render a model of a human face because of a desire to smooth the appearance of the different curved surfaces.

Surface Highlights. Surface properties provide a viewer with crucial information about the constitution and identity of an object. Surfaces can be described as dull, shiny, rough, for example. Subtle cues, such as reflections and highlights, allow the viewer to infer a great deal about a given object. These same kinds of cues can greatly enhance 3-D models.

Surfaces vary considerably. Some are completely smooth and shiny, like those of a new car; others are dull and nonreflective, like those of a carpet or rug. *Diffuse reflections* appear dull and flat the way an object painted flat black does. A flat black object reflects very little, and so the source of light is not discernible. The diffusion effect is caused by rays of light reflected at random. On the other hand, if you were to take an ordinary piece of plywood, finely sand it, and apply a glossy varnish to its surface, specular reflections would be apparent. *Specular reflections* always indicate the origin of the light, that is, the relative placement of the light source. Mixing diffuse and specular reflections allows you to create the appearance of realistic surfaces and enables the animator to suggest any kind of surface texture.

Many rendering programs provide a considerable range in variation by adjusting two attributes: the intensity of an object's reflection and the width of its reflection. The difference between dull and shiny is the difference between two qualities; the intensity of the dullness and its width can be made to vary simultaneously from all the way "off" to all the way "on." But what happens when the width is turned all the way on and the intensity is all the way off? In such a case, an object would appear plasticlike, which may or may not be appropriate to the object you are depicting. With the dullness and width turned all the way on, the object may appear too reflective, like highly brushed steel. Using these two attributes, you can cause the surface to mimic characteristics of real materials (see Figure 9–3).

Some rendering programs allow you to shift individual colors with regard to their illumination from the light source, in much the same way that you would expect to see when viewing well lighted highly metallic objects. If you observe closely you will notice that many plastic objects retain their color when a light source is moved. This is not, however, true of certain metallic objects. For example, think of an object, such as a cooking pot, that has copper surfaces. Copper's color is dependent on the angle of light. This quality of many metallic surfaces is sometimes referred to as the *metallic coefficient*. Some software programs provide you with an option whereby you can select the specular attributes of

How Smooth Is Smooth Enough? A smoothly shaded object such as a dome might appear to be properly shaded and lighted when rendered, but it may appear faceted when it is animated. When the light source goes over each separate polygon, the point of the highest reflectivity may stand out and inform the viewer that the sphere is not really round. If you require a perfectly smooth surface for an animation, check out its appearance when it is animated. If it appears faceted, then the object will require either more polygons or less reflectivity. Reducing the reflectivity will make the surface appear more dull. This kind of unwanted effect is especially visible when the object in question is viewed close up.

particular metals or surface types from a listing of specially configured files.

Transparency. An effect that brings an added sense of realism to a rendered image is *transparency*. Of course, a fully transparent object can't be seen, so transparency is assigned to objects in varying degrees. The appropriate value often must be arrived at by trial and error. However, when you are familiar with the ways in which an object's surface characteristics affect one another, you will find that there are predictable nuances of which the experienced animator can take advantage.

You should keep in mind that a transparent object (and its separate polygons) retains all of its surface characteristics even when it is assigned complete transparency. For example, a transparent sphere

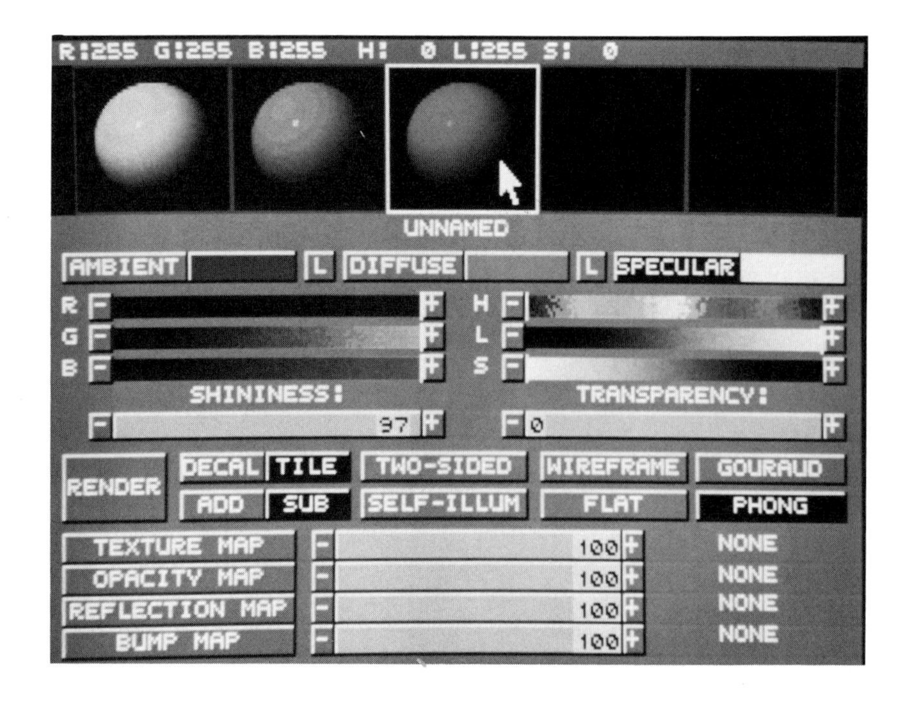

Figure 9–3. *A material editor provides a means for creating and testing different colors, qualities, effects, and maps on a sphere. (3D Studio software reproduced with permission of 3D Studio.)*

whose surface has been assigned as Phong having high reflectivity will reflect only its highlights, and so it will appear much like a glass bubble suspended in mid-air. This, it turns out, can be an extremely felicitous outcome: You can use this effect to craft realistic glass windows on buildings. Virtually transparent windows reflect only the light sources and so can look deceptively realistic.

Making Clouds. One particularly useful application of the transparency effect is in the generation of synthetic clouds. Here's how it's done. First, create several spheres and scale only the x-, y-, or z-axis so as to create a number of different eccentric blobby shapes. Next, assign each of these forms a slightly different level of semitransparency and grace it with slightly off-white color. Then assign their surfaces as Phong, and make them diffuse-shaded. Next, position the objects so that they are close to one another and overlapping. Finally, render them. You will find that these synthetic clouds are not true to life, but experimentation will usually provide you with adequate stand-ins for most production purposes.

Because most starting animators are not used to working with the variable of transparency, it is a good idea to experiment with the transparency function of your rendering program. You might, for example, fashion a block and assign it semitransparency. You will discover that, indeed, all sides continue to evidence reflectivity based on the available lights. Next, designate only one side as reflective to make the object appear more natural. Also notice the effects of color. If you select a red light source and the block is green, for example, your object's sides will appear yellow—perhaps an unwanted effect.

Special Effects Surface Tricks. In motion pictures, black regions are used as place holders on the original film in order to provide for the inclusion of special effects or live action at a later time. The technique is referred to as a *traveling matte* or *hold-out matte* (see Figure 9–4). By way of a traveling matte, a synthetic object can be created that will obscure objects that would otherwise appear behind it. Later, other image elements designed to fit into the matte region can be re-exposed onto the film. With video, the same process of creating mattes is used when complex visual elements, such as live action and computer generated characters, are composites.

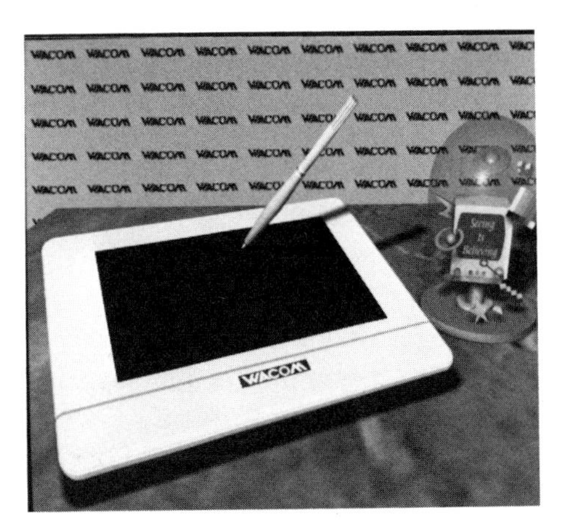

Figure 9–4. *A traveling matte is useful for adding special video effects over a particular object. The tablet surface can be assigned to be a video image replacing its matte. Later, the 3D animation and the video are combined into a perfectly fitting composite.*

Texture Mapping. Surface characteristics assigned to an object can provide a impressive sense of realism. Other techniques can be even more effective, however. One alternative is to superimpose a completely independent image over an object; the image becomes, in effect, the object's surface. This technique, called *texture mapping*, is a powerful tool that allows the animator to imbue an object with a remarkable degree of realism. See Color Plate 5 for some examples of texture mapping.

Texture mapping can dramatically modify the surface characteristics of an object. The process adds vitality to objects and provides the visual cues that allow a viewer to project character onto an object. There are two different approaches to texture mapping. You can modify the surface look by using 2-D image manipulation, or you can modify surface characteristics in all three dimensions.

The simplest 2-D texture mapping method is called *projection mapping:* A 2-D photograph of wood grain, for example, is projected onto a surface in much the same way that a slide projector would superimpose an image on a wall. If the surface of an object (a wall, for instance) is uneven or somehow irregular, the projected image will warp accordingly (see Figure 9–5). In the same way, a texture map is simply projected onto the surface of a polygon. Furthermore, the map can be adjusted independently in the x or y dimension to fit the contours of a given object.

This kind of mapping process is called *parameterizing* the surface of the polygon. In most cases, the parameterized object's surface color remains unaffected, even though the map may be colored. Thus the object's original color may or may not be visible, depending on the intensity and color of the superimposed map. In fact, many programs allow the animator to control a map's level of opacity. This feature allows you to dictate the density of the map that is projected onto an object. Using a low level of opacity with a wood textured map, for example, would only partially occlude the surface of the underlying object.

Another texture mapping method is called *wrapping*. Here an image (say of wood grain, again) is "wrapped" around a selected object

Figure 9–5. *The texture map in this image was a 2-D image of a grid. (TOPAS software reproduced with permission of AT&T Graphics Software Labs.)*

and thus exactly envelopes it. In this kind of mapping, the sides of a texture wrapped object appear the same as does the facing surface, and thus they evidence no image distortion.

Wrapping is typically used when an image (of the world, for example) is to be superimposed onto a sphere to produce a global effect. It's important to note here and in other mapping situations that the ends of the map must appear to meet seamlessly. If a design is to be wrapped over a sphere and care is not taken to match the ends of the image, the viewer might be able to detect the seam on the mapped sphere where the left part of the 2-D map meets the right side of the mapped image. The 2-D editing process is done independently on the 2-D paint program.

In such a case, a paint system is used to make the ends look the same by copying one side to the other. Smoothing or painting over the seam will then make it disappear into its opposite edge. However, before repair work is needed, it's best to check the mapping program to gauge the dimensions of the required map. These may not be easily accommodated by the paint system screen. If conserving memory is a concern, several maps can often be stored together as one digital image so that you can select part of the image to impose on different objects.

Another type of map is useful, dramatic, and effective for propagation of realistic imagery. In *reflection mapping* (or environment mapping, as it is sometimes called), an object is made to reflect a facsimile of its surroundings like a mirror. In Color Plate 3, for example, the glass table-top is reflection mapped. Whether your software calls this process reflection mapping or environmental mapping, the process is the same.

Here's how it works. First, the program must render what the surface of the object "sees." That is, the program precomputes a rendering of the surrounding objects based on the point of view of the reflecting object. This rendered image is naturally misshapen because the computation is based on a viewpoint as seen on a sphere; the result is similar to the effect of looking at a mirrored Christmas ball. Thus a rectangular-to-spherical transformation always appears distorted. The result of this rendering is automatically saved in the texture map buffer (RAM or the hard disk).

Next, the image is rendered normally but the object possessing the reflection map uses the first rendered image as its texture map. The rendered image of the other objects appears on the reflection-mapped object, making it look like it was made of chrome. This also means that moving objects in an animation will be seen as moving too.

As you might imagine, a few difficulties are associated with this process. Because the texture map of the reflection is incomplete (for example, double reflections of itself are missing, and errors arise in interactions between creating objects) the image is not truly or mathematically correct. You also need to be aware that the visual complexities of semitransparent objects can generate other visual errors. For example, if a semitransparent reflection mapped object is to display the image behind it, it depends on what was rendered first, the object or its background. Still, most of these errors can be ignored because the overall effect is often more important than image integrity. In many cases, even the most observant viewer will not have enough time to catch these minor errors, and so there's little reason to attend to them. Keep in mind that the time required to render a reflection map is nearly double that required to render the reflected image alone because the reflection map is created in addition to the requisite rendering of the environment map.

Reflection mapping can also incorporate texture maps. For example, you could take a map of clouds and then combine the reflection and the clouds together into one map. With some experimentation, this mapping technique can be an extremely effective visual tool.

Bump Mapping. *Bump mapping* is method of generating surface texture that's in a class by itself—it alters the appearance of an object's surface characteristics, not just surface appearances. Bump mapping affects surface orientations such as bumpiness, whereas texture mapping affects only the reflectivity of the surface. A bump map takes a 2-D image and changes the characteristic appearance of the object's surface in all three dimensions. It's important, however, to emphasize that it changes only the appearance of the object—the 2-D map changes only the way the 3-D object looks and does not change the 3-D model.

When you bump map, you alter the amount (or height) of an object's surface displacements: The degree of permutation is controllable and must be specified. A classic example of the use of a bump map is the creation of an orange. Although a texture map could be used to wrap an image of an orange around a sphere, the shadowing would be static and lifeless. Think about it. How would an animated orange appear when a light source changed direction and the shadows then seemed as if they were headed off in the wrong direction?

Even though bump mapping is a powerfully realistic tool, it is limited—it can affect only the shading characteristics of a surface, not its geometrical shape. For this reason, it is sometimes called *normal perturbation mapping* because it changes (perturbs) the face (normal) of a polygonal object. If you look closely at an object that has been bump mapped, you'll notice that the edges perfectly match the object's shape: They don't shift to reflect irregular surface characteristics (see, for example, Figure 9–6). Still, bump mapping is often the animator's tool of choice because it is much faster to bump map than it is to change the actual 3-D characteristics (dimensions) of an object, and 3-D manipulations, of course, take much longer to render.

An alternative to bump mapping is *contour mapping* (sometimes called displacement mapping). When you contour map, you change the actual geometry of a surface. Rippling water, for example, can be created via contour mapping. The contour of the water is created first, and then a surface sine generator (a wave maker) is used to make the ripples. Finally, the repeated image is made to ripple periodically. This technique works well when irregularities are relatively limited, but problems can arise when many contours need to be managed or when the contours are very large. You may find that the map leaves gaps between polygons or that parts of the object interfere with other parts.

Many animation systems contain a library of maps and/or images. You'll usually find that a small collection of texture maps, bump maps, and displacement maps has been included with your software. Some of these maps might include such effects as marbling, wood grains, random patterns, waves of various sorts, and metal foil.

a

b

Figure 9–6. *Bump maps artificially add a quality of 3-D "bumpiness" to objects. The block in (a) was assigned a bump map and a texture map of an image of wood (b). (3D Studio software reproduced with permission of 3D Studio.)*

In order to conserve memory and still provide access to maps, some programs can generate custom maps mathematically, based on criteria prompted by the user. In this way, you can decide, for instance, just how marble should look. For example, the marbling can be made to appear densely colored and streaked, or the coloring can be only sparsely striated. Other constraints include definition of bumpiness, as well as the size and/or roundness of the map. With these controls you can fabricate objects that are incredibly unlikely or difficult to make. See Color Plate 6 for an example of an animation that was created with the help of many of the techniques discussed in this chapter.

Overall Quality Control

Now that your objects have been forged in their entirety, you are ready to introduce them to the 3-D stage, which includes the dramatic addition of cameras and lights. You can garner additional control of overall quality of rendering via software controls that add a measure of precision to the rendering process. However, the price for higher quality is, as you might expect, additional rendering time. You will discover that, in many cases, the enhanced quality of the output is worth the extra time and effort. Software quality controls can provide you with

- overall rendering quality with anti-aliasing
- overlapping of objects
- sharpness or dullness of the light reflections

For situations in which a fast turnaround is needed, some programs provide a shortcut to anti-aliasing, which is a technique that blurs the edges of objects. Although the shortcut reduces rendering time, animation professionals would quickly spot the difference.

Lights, Camera . . .

10

One of the most gratifying steps in the animation process is to behold a 3-D model that has been transformed into a vibrant, vital, image. This transformation is accomplished at the end of the rendering process, when lights and a camera's viewpoint are added to an image to make it "picture perfect." The parallels between this step and the making of a feature film are noteworthy: At this point in the production you assemble the actors (in full makeup and costume) on stage and run through your scenes with the lighting director and the cameraperson.

The cinema analogy breaks down, however, when you compare the production methods in relation to time. It can take from a few seconds to as much as a day for the computer to render a single image from the digital model (although a few minutes is about average). This is especially considerable when you recall that even a simple one-minute animated video sequence will consist of 1800 images or frames. In fact, many professional animation facilities consider it unprofitable if rendering requires more than two minutes per frame.

Along with being expensive with regard to time, the rendering process monopolizes one or more computers until all frames have undergone processing. Waiting for an animation sequence to be completely rendered can take days. It seems that no matter how long it takes (rendering time is a function of a computer's overall speed), it's always too long. This animation fact of life is one of the main reasons why there's continual demand for faster computers and software in the world of computer animation.

This final rendering process is particularly exciting, but it can be frustrating for otherwise highly skilled animators. Model-making skills place a premium on procedural kinds of attributes—it's an exacting and tedious business. However, this is where the animator changes hats. At this point, aesthetic considerations come to dominate the animator's decision-making process. You will quickly discover that choices about lighting and shadowing can transform a mundane title sequence into a spellbinding miniature production. The effects of your artistic preferences can culminate in a sublime overall appearance to your work or in a visual debacle.

To realize your visual intent through the rendering process, you need to have a thorough understanding of color, surface properties (covered in Chapter 9), and the effects of lighting. You'll find that, when you get to rendering, you'll become intimately acquainted with the capabilities as well as the limitations of your system. You're sure to run into some surprises. For example, some programs will create shadows only on certain kinds of shaded objects and will not make shadows from semitransparent objects at all. Some programs create shadows with so much granularity or other unwanted artifacts that the shadows can't be used unless they are moving too fast for adequate scrutiny. You will eventually learn to work with your animation program's rendering capabilities in the same way that artists learn to work with their tools: Used with care and forethought, your tools are usually adequate for most jobs.

What the Software Does

After a 3-D model is created and all of its surface properties have been assigned, you process a fully rendered image by providing

- the position and colors of the light(s)
- details regarding shadows (such as position of light source, and color)
- a virtual camera position and lens type

To fully utilize the interactive features of rendering software it helps to have an overview of what the process entails.

Translating a 3-D model into a 2-D image is a complex and computationally intensive process. The state of the art is constantly being challenged by new methods. Developers of animation products select from the latest and most innovative software algorithms and incorporate them into their latest versions. Software can make a decided difference in overall rendering speed, as you will quickly discover if you try using different software on comparable—even identical—hardware.

Most rendering programs take the 3-D model and scan for visibility first—that is, which object surfaces are in the foreground. Next, the program typically calculates how to light the model according to its surface properties (dull, reflective, and so on). Perhaps the surfaces are to have a 2-D image superimposed. In such a case, the distortion of the angle of the image must be calculated.

The rendering program must also calculate how the scene will be viewed, that is, how it is to appear from a particular imaginary camera's point of view. If shadows are to be calculated, all parts of the image must be assessed. Does a certain midground object cause or receive a shadow? How should the shadow look? Once these aspects have been determined, the program processes the image pixel by pixel and one horizontal line at a time, and you will see the completed lines slowly appear on screen. The rendering time required is a function of the number of polygons composing the model and how many features (such as transparency, shadows, and lights) have been enabled.

Lighting. Although lighting is integral to the classification of objects (covered in Chapter 9), it warrants more concentrated discussion here. Indeed, control of lighting is what makes for the "artistic touch" that transmutes the mundane into the extraordinary. Subtle changes in lighting can have fascinating effects on a synthetic image. The ability to define the number, color, and positions of lights allows the animator to control the overall appearance of an image. Nuances of lighting can convey a wealth of information to the viewer; and even slight changes in lighting elements can have profound effects on the image's impact.

Lights make a statement about a scene and about the intentions of its creator. As you design your animation, think of yourself as the lighting director of a major film. A good lighting director knows that the position and intensity of each light does much to control the mood of a scene: Lighting choices cannot be casual or arbitrary. See Color Plate 7 for an example of lighting used to control the mood of a scene.

Individual lights possess any or all of the following attributes:

- color
- position
- intensity
- direction

Given a specified amount of ambient light, luminous positions may be set up in a 3-D space as point sources. What's more, depending on the software in use, color assignments can usually be given to these light sources. In some software packages, lights can be sized so that they only illuminate or "fill" certain areas. Other animation packages provide you with a completely artificial light source, such as a spotlight, that acts more like an object than a light. This light source may be completely animated as if it were a floating light bulb traversing a scene. Keep in mind that as you add light sources to an animation, rendering time increases—each source will require additional calculations.

An important consideration is the proximity of a light source to an object. A light illuminating a certain area will affect other nearby objects in a similar manner. Often you'll find that placing lights far from an object makes for a more "normal" effect.

Some applications call for the lights to be positioned exactly as the sun appears at a particular day, time, and location. Incidentally, such lighting verisimilitude is a requirement in the planning departments of some cities. Such accuracy can help planners visualize how a proposed building's location will affect its surroundings. Will the building, for instance, severely overshadow neighboring structures? With most rendering programs, there's no need manually to calculate the position of the sun using your (probably dormant) geometric and trigonometric skills—the software provides this service for you.

A heliometric option on a rendering program provides you with this capability: All you do is enter a time of day, a geographic location, and the light source's position, and the program will calculate and cast the appropriate light (and shadows) as would the sun. Animating these representative positions of the sun (as you will see later) is also easily managed.

Depth Cuing. One way to add photo-realism to a scene is through the use of *depth cuing*. In this process, objects are vested with more "background" color the further removed they are from the foreground. Some programs allow complete control of distance and color values so that objects that are very far away are seen to disappear; that is, they appear to merge completely with the background.

A similar effect can be achieved through an effect called *fog*. Fogging allows you to mix a particular color into a scene. Because the density of color infusion is graduated—color is less saturated with distance—objects exhibit less contrast (even though they remain in focus) the further removed they are from the light source. Color Plate 8 demonstrates effective control of lighting with regard to distance from the viewer.

Here's another depth cuing trick. Some advanced renderers provide so-called *dark lights*. As you might imagine, a dark light is the opposite of a light: It radiates darkness over particular areas and objects. This effect can be very useful when you want to add a touch of drama, for example, when you need to reduce a particular object's ambient light so that it can be effectively lit from underneath.

Another useful feature found in software packages enables you to light a specific object while withholding lighting from surrounding objects. This feature may include the additional provision of control over specular/diffuse lighting conditions for each object.

Shadows. An adjunct to the lighting determination, and a critical visual cue of realism, is the addition of shadows. Skillful shadowing can imbue even the simplest image with startling realism. As a technical matter, however, effective shadowing can be hard to simulate despite its seemingly intuitive nature.

Software that automatically generates shadows is based on the directional pattern of light falling on an object coming from a particular light source. A calculated image is mapped onto an object, creating the approximation of an actual shadow. To perform automatic shadow generation in the context of a lighting environment you need to know about the various attributes that describe shadows. The primary qualities of shadows are

- edge thickness: the sharpness of the shadow's edge
- granularity: is the shadow smooth or grainy?
- model calculation size: controls shadow quality
- scene integration: how other lights interact
- rendering time: shadows always take longer to render

The implementation of these effects varies considerably from one software package to the next. Shadow effects are individually controllable with the more capable animation programs. As a general rule, however, better the largest shadow model size (as measured in pixels) achieves shadows of the best quality. Shadow model size is often determined by the user as one of the available options. As always, there is a price to pay: Larger shadow models require more memory and take longer to render.

In order to render shadows, internal calculations are made at a particular resolution; the higher the resolution, the better the overall quality of the shadows. A high-resolution temporary digital image consumes a significant amount of memory, however, and you may find that your computer (or program) is not able to produce the level of quality you require. For testing purposes to assess the overall quality of your shadows, a low-resolution shadow map can provide you with an idea of how the light source will project shadows.

As you evaluate a program's ability to generate shadows, experimentation is in order. Each light (including ambient light) interacts with other lights, and this affects the shadow. One quality to pay particular attention to is granularity (dots of blackness instead of smooth gradations of darkness). An image may appear acceptable as a static image, and yet it may not work as a moving image. If granularity is too high, so that the shadow is composed of coarse black dots, each rendered image will generate shadows composed of dark dots in a slightly different position, causing the shadow part of the image to scintillate noticeably during animation. The eye can perceive these differences between each image in an animation; the effect is distracting and almost always unacceptable. Other shadow artifacts, such as patterning effects (for example, Moire patterns), may also appear only during animation. See Figure 10–1 for some examples of shadow settings.

Due to the added time it takes to render shadows, many animators (including Hollywood's best) fashion fake shadows with flat-shaded polygons. The major benefit is that rendering time is, as you might expect, reduced dramatically. However, this kind of treatment should be attempted only in situations where there are relatively few objects. The eyes are easily fooled, but they are quick to spot blatant incongruities. You'll need to depend on your artistic skills to engineer the proper positions of the counterfeit shading.

Sometimes the shading effect can be set up easily, as is usually the case with shadows for text (called drop shadows). Text is first created and then duplicated. The duplicated text is made to appear semitransparent, nonreflective, and flat black. It is then offset behind (in the z direction) the original text. Assigning the shadow text to act as a subordinate of the readable text will make the shadow move perfectly with the motions of the text. However, if the background is complex, viewers may quickly—and often unconsciously—notice that something is wrong, and so the trick fails.

Figure 10–1. *User-adjustable shadow settings affect the final appearance of objects, such as their granularity and sharpness, in addition to affecting total rendering time.*

Camera Point of View. This section looks at the impact of another influential player on 3-D scene, the virtual camera. After 3-D objects have been created in an explicit 3-D space, the scene must be viewed from a distinct *point of view* (POV). What's more, the camera's POV must also be imbued with the qualities of a virtual lens, much like the lens of a camera. That is, the angle of vision, wide-angle or telephoto, must be defined.

Two popular methods are commonly used to control the camera. In one method, the camera's POV is what you see on the screen. The view position, angle, and so forth can change accordingly. The problem with this method is that the viewpoint can sometimes be difficult to imagine without other visual cues. A common software aid used to help you observe and clarify camera position is the provision of multiple simultaneous views; for example, top, bottom, and side views are all provided on one screen (see Figure 10–2). This type of interface is sometimes called a 3-view. Using the information provided by a 3-view, the animator can get a better idea of where the camera is in relation to the animated objects.

The other method commonly used in animation applications is to treat the camera as if it were simply another object within the context of the animation. Given an already defined 3-D space, a cameralike object can be made to mimic a camera whose position is readily moved about within the scene. This camera's rotation, lens, and other attributes can be controlled and edited in much the same way that other objects in the animation can be controlled. Furthermore, its path can be marked (by dotted lines, in some software) to indicate its position relative to surrounding objects. In an on-screen window, the software typically displays a representation (often in wire-frame) of what the camera is seeing. This feature allows the animator/camera operator to control and verify the elements in an animation so that a good idea of the overall effect can be envisioned before it is rendered.

As you might expect, much of the terminology of computer animation camera movement is drawn from the language of filmmakers. Here

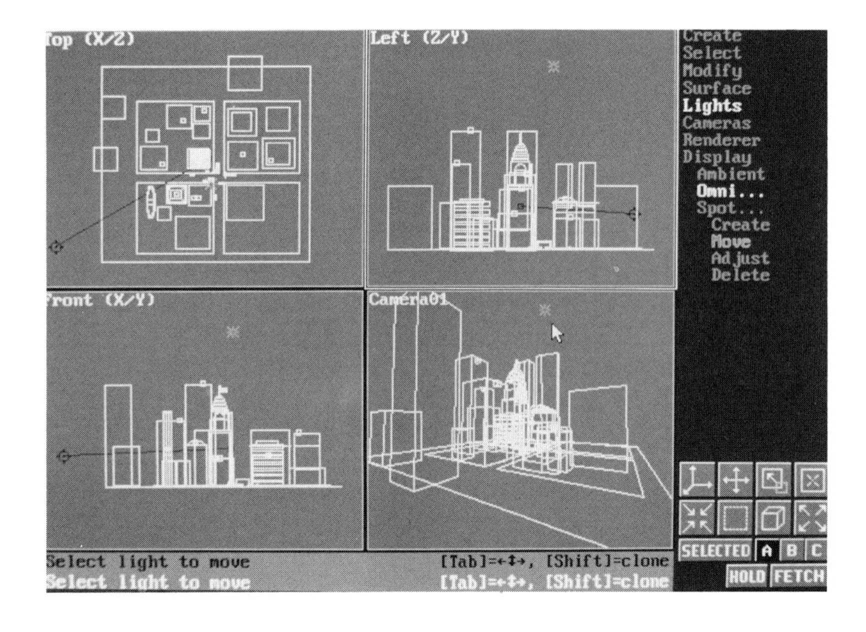

Figure 10–2. *The camera's position can be moved easily. Simultaneous multiple views indicate the location of the virtual camera.*

are some of the common terms shared by cinematographers and animators:

- *Panning:* the process of rotating camera POV about a vertical axis
- *Tilting:* similar to panning but produced by pointing the camera up and down on its axis
- *Zooming:* a movement similar to dollying except that the camera lens' focal point is changed
- *Trucking:* a sideways camera movement that parallels the movement of an object.
- *Dollying:* moving the camera toward or away from the objects in a scene

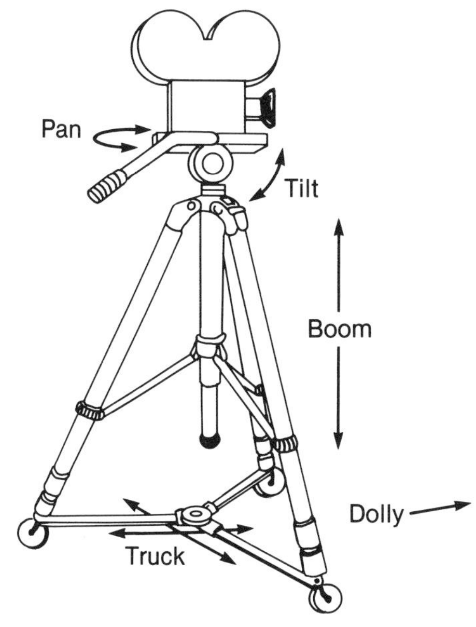

Figure 10–3. *Conventional cinematography terms are also used in digital animation, with a virtual camera.*

Figure 10–3 shows a diagram of an actual camera with the camera movements indicated.

Finally, the camera's perspective can also be changed; this is similar to changing the focal length of the lens on a camera. The focal length is determined by the distance between the viewing window and the POV. You can adjust focal length to produce a wide-angle effect or a telephoto effect. Keep in mind, though, as you zoom into a view, your focal length changes. That's why you need to know whether you want to zoom into an object or whether you simply want to change your POV (where you, the viewer, are standing relative to the object).

As you experiment with the POV and perspective features of the virtual camera, you will realize that you can create dramatic and bizarre effects. You will find that a software lens can far exceed the capability of any real lens. For example, if you so choose, you can position the camera an inch away from a building and then adjust the perspective so that you can view the foundation and the rooftop at the same time. This is an excellent way to enhance visual drama in an animation. It is also a popular cinematography technique and is called *forced perspective.*

Graphics Standards: The RenderMan Interface. Because of the increasing complexity of today's software, a new technical terminology is evolving, one that is often redundant, inconsistent, and even outright misleading at times. Confusing terminology continues to be a problem for the animator. In many instances, we still have no reliable and consistent way to describe the essential elements of a scene so that it can be dependably managed by the rendering components of an animation system.

After all, a completed animation represents the accumulation of numerous subtle decisions, choices that can easily be confused in translation from one software module to another. That's why there have been several attempts at the creation of a standard language interface. One notable example, from Pixar, is the RenderMan interface, which addresses most 3-D rendering attributes. It offers consistent technical terminology and is, in effect, a fully realized description for rendering.

Pixar's development goal was to separate 3-D software functions so that a common description language could pass data and instructions from the modeler to the renderer. Another goal was to facilitate hardware interfacing—to the extent that it was in compliance with the program's methodology. Thus, the RenderMan interface is a way of describing the details of a scene so that a renderer can accurately and predictably do its job.

For example, if an environment map is to be used, the language provides a consistent way to manage the different sets of variables (such as camera viewpoint and quality of resolution) that allow the animator to determine how the scene is to be created and how it will ultimately appear. The result is that the animator gains more fluent control over the details that govern how an image is to be rendered.

RenderMan addresses functions that are similar to those associated with basic primitives. Its capabilities include hierarchical modeling, camera modeling, shading attributes, and control of many other rendering-related features.

Use of RenderMan in animation production is already extensive, and even inexpensive modelers such as the Amiga systems are now utilizing a RenderMan interface. These kinds of systems are set up so that they can be used to render the images directly, or, if the animator so chooses, the rendering data can be ported to a more powerful system that supports the interface. Thus, the standard provides a communications bridge between different desktop platforms and animation products.

Finally, a few words of caution are in order. RenderMan does what it sets out to do, but its capabilities are limited. It does not provide animation control; nor does it serve as a motion-scripting language. It's outstanding utility is as a rendering interface after modeling has been completed.

From the standpoint of the animator, RenderMan is a significant step in the right direction, that is, in the direction of a more standard relationship between modeling and rendering. Animators who choose to use the RenderMan interface need first to make sure that their animation software supports the interface.

Adding More Realism: Ray Tracing and Radiosity

A historically old method of rendering an image is called ray tracing. As its name implies, *ray tracing* is based on tracing light pathways from their source. A ray-traced image is created by accurately following the course of each light ray as it travels through a particular scene. The result is excellent photorealistic images that project perfect shadows (see Color Plate 9). The primary drawback to a ray-traced image is that it takes an exceptionally long time to render—sometimes as long as a day, depending on the complexity of the image. A ray-traced image can accurately depict shadows, reflections, refractions, transparency, motion blur, and multiple reflections of objects on other objects.

An explanation of how it works may help you to understand its virtues and downsides. Ray tracing was one of the earliest methods used as a rendering technique. The method is based on the physical laws of optics and light—the same laws that describe how a lens works. The process works by following the path of each individual ray of light coming from a light source. A light ray travels in a straight path until its route is interrupted by an object (thus illuminating it). This same light ray may bounce off the object (depending on its reflectivity) and illuminate some adjacent object.

A process that followed each light ray from its source would be wasteful because most of the light rays never reach the viewer. That's why ray tracing works the other way around, starting from the viewer and tracing a path of light to its source as it bounces off objects. In this way, all the light can be accounted for in an efficient process. The result is that objects are illuminated in an optically accurate and realistic fashion, inclusive of shadows. Despite impressive image quality, it is one of the slowest rendering methods.

Ray tracing has other problems. Sometimes, as in real life, the calculated ray of light will pass through transparent or semitransparent objects, and in so doing it will cast an image on the surface as if it were a lens. The resultant artifacts are called *caustics*.

Another problem comes about as a result of diffusion. For example, consider light reflected off a shiny blue object and onto a second object nearby. You would expect the second object to evidence a diffuse bluish cast to its reflected light. Unfortunately a ray-traced image can not produce this realistic effect. Instead, second-hand reflections are handled by another technique called radiosity.

Radiosity also imposes a stiff rendering-time penalty but produces high-quality images that accurately portray lighting effects in situations where you would expect the lighting of one object to impact another. The results can be startlingly realistic. A technique called *diffuse interreflection* manages the lighting relationships of objects relative to other objects on a point-by-point basis. The result is accurate shadow descriptions including such details as preumbras and penumbras—a truer simulation of reality.

Finally, you need to understand that there is no one perfect rendering solution to a given image or set of images. There is no "best way" to render. Considerations are based on two determining variables: time and quality. Perhaps the best program is one that evaluates what has to be done and uses the appropriate algorithm to execute it. Not surprisingly, this is exactly what is done in the best research facilities and at the better production facilities. Although this capability is presently out of the range of PC-based animation systems, newer and more intelligent programs are likely to be available in the near future that can make these assessments.

It should be noted here that photo-realistic images can either be rendered on the display system or saved on the hard disk. In fact, images may be rendered at a resolution higher than the graphics system's capability. Resolutions above 4096 horizontal lines can be saved on the hard disk, which can then be sent to a device called a digital film recorder and later processed as high-resolution slides. This capability is covered in greater detail in Chapter 15.

. . . Action!

To continue the feature filmmaking analogy, imagine that your actors (3-D models) are on stage and in costume. The lights have been carefully set up, and the camera is positioned for the opening scene. You now need the script that defines the action for the actors and the stage hands. This script is a plan of actions with reference to time and is used to guide the action of the 3-D animation just as a conventional script guides conventional stage and live-action motion-picture productions.

Animation programs handle the motion-scripting process in different ways and use differing conceptual metaphors to describe and manage the process. Some animation programs actually refer to "actors" and "scripts." Regardless of what methods or terminology you and your program use, watching your images come to life and move for the first time is indescribably exciting. To some extent it's an excitement that will ultimately be shared by viewers of the production, who will have an opportunity to experience your vision.

All animation is based on the illusion of movement. In a conventional (noncomputerized) animated production, motion is simulated by a succession of images, which are created a frame at a time. This chapter focuses on the frame-to-frame changes that create this illusion of action. The focus here will be on the use of 3-D objects, but processes we look at apply to 2-D objects and animation as well.

You will find that articulating the motion of objects can be a laborious and challenging process. Whereas rendering is a process of mimicking reality with visual cues, motion control is more explicit and unambiguous. You will soon discover that scripting the motion of even the simplest action can be a surprisingly difficult endeavor to manage and control. Research continues in an attempt to find better and more automatic methods for describing how to move objects in natural ways. In the meantime, our current tools provide us with the ability to turn our "real" 3-D objects and scenes into action.

Getting Started

The process of animating the players in a 3-D model involves assigning movement to the different objects, telling the "camera" how and where to look, and then previewing the motion in low resolution. In Chapters 13 and 14, we will describe the last step: how to put a high-quality version of your animated production onto videotape or film.

Because assigning motion to 3-D objects requires that the animator know the relative positions of all participating objects in all three dimensions, you need to be able to visualize a number of differing points of view, simultaneously if possible. This kind of visual overview can be presented via the software by dividing the display screen into quadrants (as described in Chapter 10). Three windows on the display screen are configured to display orthographic or nonperspective views of an arbitrary point in space for the top, bottom, and side views of the objects. The fourth viewpoint displays the camera's POV, whose attributes can also be varied and animated.

The animator can also choose to direct the motion-scripting process using only the camera's POV on a single full-screen display. This point of view is helpful, particularly when you need to see as much detail as possible and want to render a model quickly. When you are working with complex models, this method may be quicker and more responsive because there is less to display on the screen. Alternatively, you can direct the software to display what some products call a "big view." This displays the camera's POV but also provides a surrounding border so that you can monitor what is immediately around the camera's view, yet remains unseen.

The camera's POV is usually framed by a rectangle that encloses the 3-D scene. Using the big-view feature, you can be continually aware of what is about to appear on the display (you can think of this as actors waiting off-stage for their entrance cues).

Keyframe-Based Animation

The most common method used to control how an object moves from one position to the next is called *keyframe animation*. This is not only a logical conceptual tool but also a historical one. Conventional cel animators, who relied on countless painted cel visuals, started by creating key positions and emotions for their characters. It was the job of the assistant animator to fill in the details from one keyframe to the next, providing accurate, well timed movement transitions.

Generating keyframes on the computer is a similar process. When two positions of an object are designated as keyframes, the computer becomes the assistant animator and fills in the sequential frames in between. This *inbetweening process* is sometimes called *tweening* (see Figure 11–1). Object positions can be chosen, and other attributes, such as color, rotation, and scale, can be interpolated as well. Additionally, most programs treat everything as objects and provide the capability for

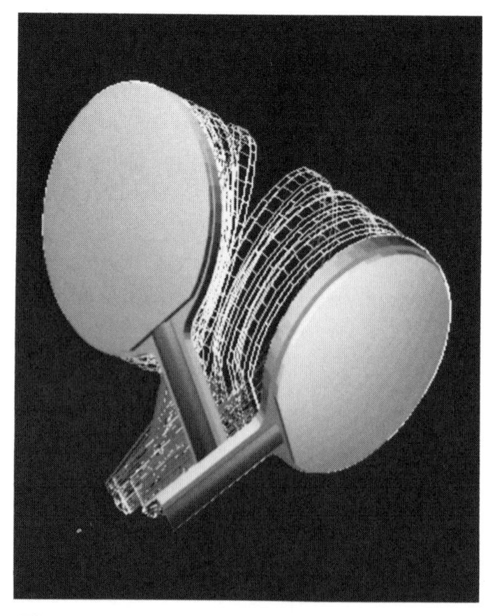

Figure 11–1. *Tweening automatically calculates all of the images in between keyframes.*

An Example of Adding Motion. Imagine you want to create an animated march of triangular objects. Your plan calls for a grid of two types of triangles: A row of 8 × 8 blue and a row of 8 × 8 yellow triangles are to be intermingled with each other. How do you make the triangles hop over each other as they would appear in a march? Scripting this action demonstrates some of the principles just presented. Regardless of what software package you are using, realize that there are several paths to accomplish this task. Here is one way of creating the animation.

Step 1. Create a triangle and extrude it into a 3-D shape having some thickness to it. Color it blue. Then duplicate it and dupe it yellow.

Step 2. You should duplicate the respective rows of yellow and blue triangles. Next, group all the blue triangles together using the first triangle as a master. Now, do the same for the yellow triangular set. You can add more realism at this point by providing a floor, adding textures to all objects and turning on shadows.

Step 3. Move the master blue triangle up from the floor — which will then move all blue triangles upward — and make this position into a keyframe. Right now it doesn't matter how much time to set the duration of the keyframe so long as it is the same for each one.

Step 4. After the first keyframe has been created, move the same blue triangle onto the ground position and make that the next keyframe. Repeat this setup, moving and keyframing the first yellow triangle. You should now have four keyframes after the starting position.

Step 5. Most software packages provide the ability to replicate parts of an animation script just as you would chunks of text with a word processor. Assuming you have that capability, copy the keyframes over as many times as you would like. If you can't replicate the keyframes, then repeat the process of moving and keyframing the yellow and blue triangles.

Step 6. Now it's time to preview what you just did. Use the preview capability of your animation program, and run the animation for real time display. If all worked well, the yellow and blue triangles should march along in groups hopping over each other. Of course, it's likely that you'll want to make changes of some sort such as in lighting, shadows, texture, or camera angles.

Step 7. Finally, after modifying the motion to your taste and verifying the results, you can now render the animation onto video tape or film.

moving and controlling the qualities of the light sources and such camera attributes as perspective and lens types.

A rotating logo provides a simple example of how keyframing works. Imagine that you have created a logo and that you want to make it appear to rotate in space. You position the logo and determine light placements and a camera angle. This setup becomes your first keyframe. Next, you rotate the logo to its rest position. This becomes the (second and) last keyframe, and you've assigned to it a particular length of time, say 30 seconds. That's all there is to it. You're now ready to preview the animation. The logo will appear in the first position, and all the in-between frames of the rotation (28 frames) will be automatically calculated and displayed.

Say that now you decide you want to move a light source. You can easily add another keyframe to the middle of the animation, and at the same time you can change (edit) any other positions (lights or camera) of the animation. You might elect, for example, to edit the last keyframe so that the camera position is further removed from the logo, causing the logo to diminish and nearly disappear at the last frame.

As you can see, keyframing is fundamental to the computer animation process. Because the computer software is capable of automatic interpolation (that is, tweening), and because the computer can store, display, and edit the activity in an animation, complex motions can be created exactly the way you envision them.

More complex model and motion attributes add to the power of keyframe-based animation. Specifying hierarchies of objects (see Chapter 8) is a feature that you can use in the animation process. When you want to control a complex animation in which objects interact predictably, hierarchies can facilitate the task (see Figure 11–2).

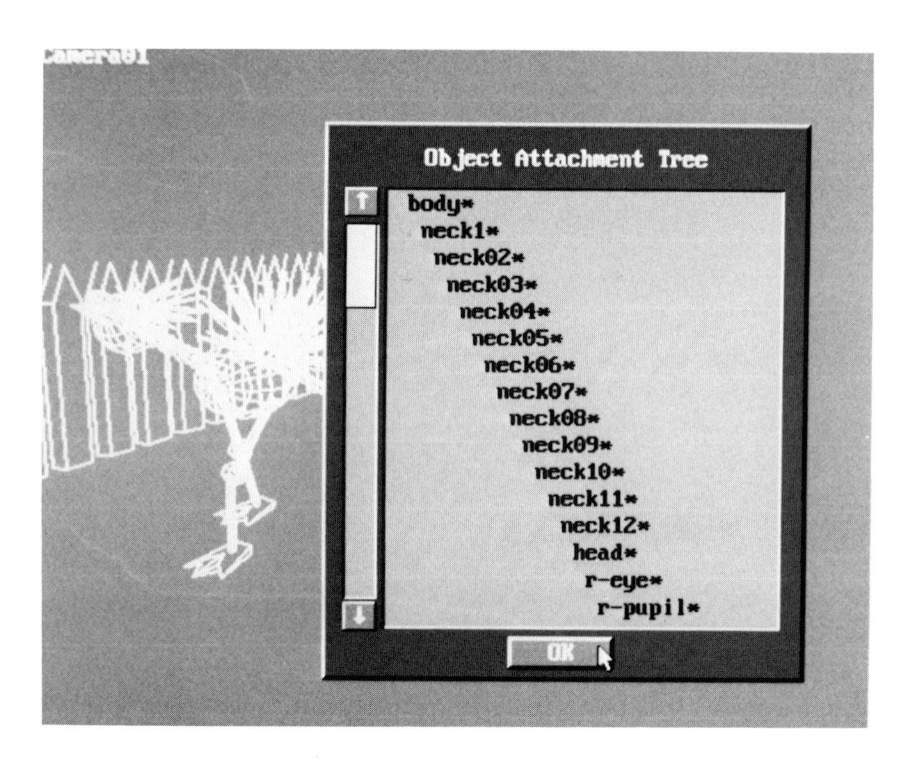

Figure 11–2. *Hierarchy offers an animator convenient control over complex motions, as of a walking bird, where the action of one part automatically controls the actions of others. (3D Studio software reproduced with permission of 3D Studio.)*

Imagine, for example, that you are animating a clock and want to coordinate the relationship of the hands' movements so they function as hands do in a real analog clock. Animating each hand would be an error-prone task as well as a time consuming one. However, if you assign the second hand to rotate 60 times faster than the minute hand and the minute hand to rotate 60 times faster than the hour hand, then all you have to do is specify the hour hand's movement—the other hands will respond according to your formulation. It's worth observing in this context that, if you so desired, you could assign eccentric movements to the second hand: after all, it has no control over the movements of the other hands because it's the lowest object in the hierarchical order. Note that hierarchies can usually be assigned, built, and edited within your program's motion-scripting functions, that is, outside the modeling program.

Additional keyframe animation capability is available via the program's editing features. Usually, the editor graphs movements in a window on screen so the animator can monitor an object's activity during a particular segment of time. This feature is interactive, so you can add, delete, move, and/or adjust the length of a particular object's keyframe in the animation. The editor also enables addition, deletion, and copying of individual objects as needed (see Figure 11–3).

Keyframe animation is not, however, without problems inherent to the process: you will discover that the higher the quality of the motion that is required, the greater the number of keyframes you must generate. That's a problem for the animator because keyframe generation is labor intensive. A high-quality (that is, realistic) depiction of even the simplest human movement requires hours of intensive labor. Nevertheless, the relationship of labor to quality is similar to that of other creative endeavors—the more you put in, the more you get out.

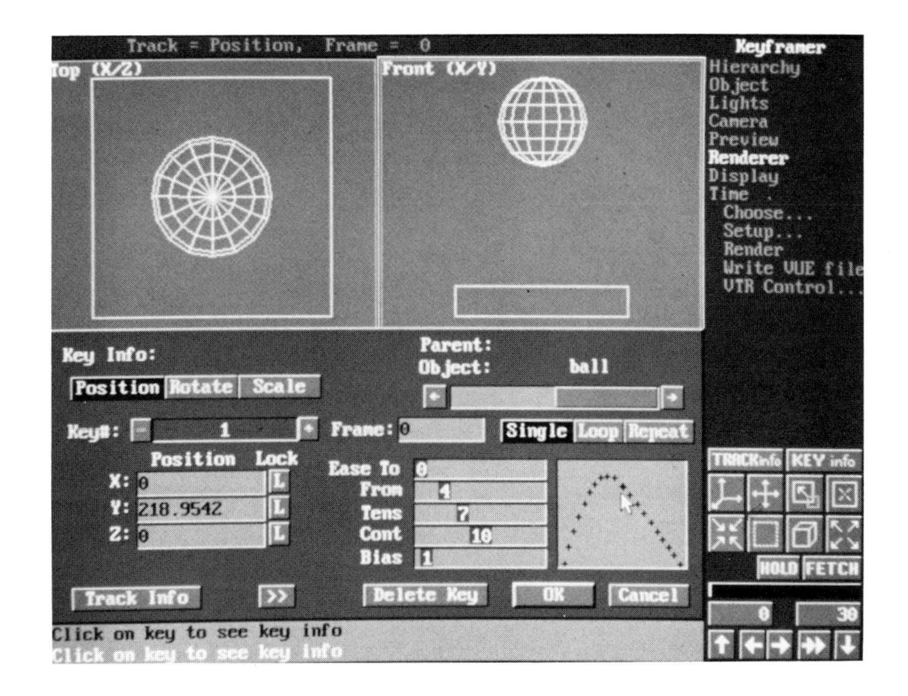

Figure 11–3. *The rate of movement in time can be controlled via a keyframe or time chart editor. (3D Studio reproduced with permission of 3D Studio.)*

Move the Camera or the Objects First? When you are scripting the motion of objects and camera, what you move first can critically affect the difficulty of the process later on. As a general rule, you should set up the motion of your objects before you create the activities of the camera. Then you should check your work; preview it to see if you are satisfied with the objects' movements. At this point, you have the option to preview the action from different vantage points so that objects don't inadvertently slide into one another. When the camera's movements are scripted first, a scene's perspective can be confusing, and this makes it difficult to move or edit objects. You'll find that at times you'll have trouble figuring out where things are.

Better Control of the Motion

When you get to the preview mode and finally see how your animation comes together, you may be disappointed if the interpolation between key frames is linear. When you have too much consistency in the rate of an activity, you'll find that the movement will appear unnatural. When a ball is dropped, for example, it does not travel downward at a consistent rate. This is why most programs provide some nonlinear interpolation capabilities. Using these features, you can make an object, such as a ball, appear to fall faster as it approaches the ground—just as it does in real life.

Nonlinearity can be quantified as a kind of curve. In fact, it is a special kind of curved line called a *spline*. Splines rely on control points, which can be used as guides or handles for the mathematical expression of smooth movements. These control points, determined by the animator, can be freely moved around to change the shape of the curve.

The simplest use of a spline for motion control is what classical animators call an ease-in. In an *ease-in*, the motion starts out slow and speeds up, that is, it accelerates. Conversely, the action can *ease-out*, or slow down with time (decelerate). This activity can be graphed with reference to time so that a quick look at the graph helps you to visualize exactly how much change there is to be. Low-cost animation programs that do not provide graphs for interactive control of speed often do provide fixed ease-in and ease-out capability.

The complex attributes of a spline can be controlled when splines are used with multiple keyframes. This allows you greater control over a spline's shape and thus more control over motion factors other than the ease-in and ease-out rate.

Due to its mathematical nature, accurate spline control is possible. Depending on the type of spline used, three primary descriptors can define a spline's shape. These are

- tension

- bias
- continuity

The *tension* of a spline detemines how accurately it meets the control points. The *bias* controls whether the curve will occur before or after the control point. Finally, the *continuity* controls the spline's smoothness. With these controls, you can create complex and highly descriptive movements—movements that would be too cumbersome to create using multiple keyframes.

Many programs provide graphs that help you to visualize a motion. These visuals illustrate a spline's appearance and graphically represent the three shape attributes. Many of these programs also provide a way of interactively changing spline shapes. Given this level of control over an object's motion, it's possible to create sudden movements, slow-downs, graceful speed-ups, and other natural yet complex forms of motion (see Figure 11–4).

Motion control via splines has limitations. In fact, if you observe closely, you'll see that sudden sharp movements in a computer animation often leave telltale spline-based "whiplash" when an object is turning quickly. You'll also occasionally see an added to-and-fro motion, which is at times nearly impossible to eliminate. The solution for exacting motion control is usually a great number of keyframes, perhaps one keyframe for every few frames. Painful and time-consuming as this solution may be, you may feel better knowing that the best computer animation facilities do it this way in order to reach their high level of quality.

Previewing the Motion

After you have assigned motion to the objects and the camera, previewing the animation in a low-quality mode allows you quickly to

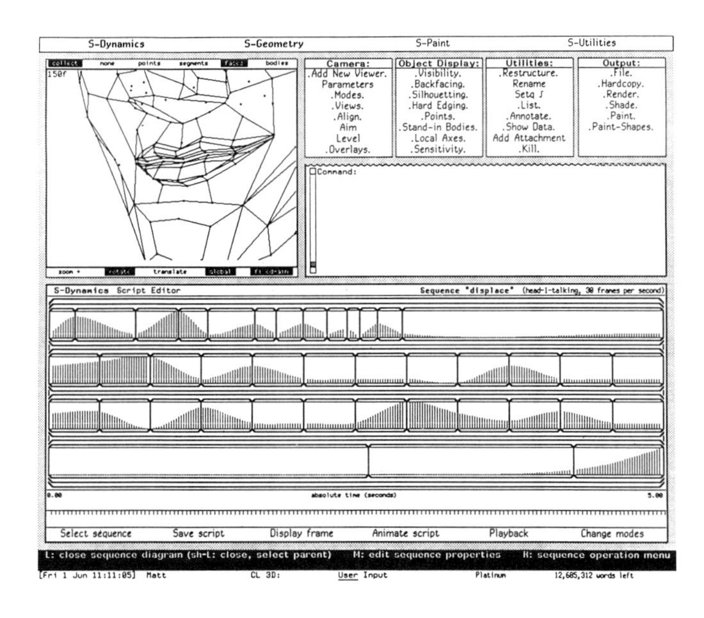

Figure 11–4. *Spline-based time charts of the motion of different objects can be edited graphically. (Software reproduced with permission of Symbolics.)*

verify the accuracy of the animation. Because animation is, as a rule, such a detailed and laborious process, editing and previewing are often repeated several times in order to fashion the most effective movements. That's why most animation programs provide several viewpoints that you can use to examine and edit the animation before it is finalized on video. The different perspectives provide you with information about specific qualities of the animation that you can use to elevate the production. Previewing methods usually include

- displaying the wireframe animation in real time
- displaying the sequence at different levels of rendering quality (resolution)
- displaying the animation with interactive manual control

The least computationally demanding confirmation of movement is usually presented in wireframe mode. Wireframe representation, like other previewing methods, allows you to view a part or all of the animation sequence. Because the memory requirement for saving and displaying wireframe images is relatively small, your animation can be displayed in real time, that is, at the same speed as it will run when the animation is finished. This kind of low-resolution sequence (sometimes referred to as a *vector test*) is typically saved in RAM for rapid playback. This method gives you an overview of the total animation experience so that you can determine whether it is too long or too short and whether the timing of the movements is suitable.

On systems that can display multiple windows, the preview mode provides a lot of useful information. For example, one window can display what the camera and viewer sees. Another can display a top-down view of a scene, indicating camera position. A third window can display pertinent time lines with keyframe indicators, as well as the amplitude of a particular object's activities—including details such as light source changes. Finally, a fourth window can display a particular object's activity so that its relationship to other objects can be viewed in greater detail.

In terms of quality, the next level of previewing abstracts the animation in a way but still provides real-time viewing. There are, however, some minor compromises. This approach creates fully rendered animations that skip a predefined number of frames. Images are displayed fully rendered but at lower than normal resolution, and the resulting sequence is saved in RAM. The more available RAM you have, the longer the sequence can be. The real-time display of such an animation sequence, even in low resolution, helps the animator to assess the quality of the motion as well as the overall quality of shadows, transparency, and lighting.

A common and useful feature of many software packages allows the playback rate to be controlled interactively via a graphics tablet or mouse. Some systems provide an additional display (on the screen or on the text monitor) of a time graph that shows what frame is currently being displayed. Such a method is shown in Figure 11–5. Individual items (such as the rotation of an object, the color of a light, or the

Figure 11–5. An animation can be previewed via two screens, one that displays the time chart and the other that displays the particular frame. The mouse controls what frame is being viewed. (Crystal 3D software reproduced with permission of Crystal 3D.)

camera's position) can also be chosen and displayed by themselves or in a column. This provides an excellent way to monitor and control an activity in relation to other activities. It also helps you view interactions between fast-moving objects, such as the details of a collision.

You may want a ball, for example, to bounce in a particular way at a specific location. Real-time previewing may execute too quickly for you to verify these details. However, gliding the stylus across the tablet controls the individual frames. Thus the speed of your hand can control the movement, speeding it up or slowing it down.

It's important to realize that this previewing process does not completely prove an animation's quality—it's only a rough approximation of the finished motion. Such problems as shadow defects are often subtle, and they are not apparent in the previewing process even though they can be a noticeable problem in the final animation. Likewise, fine details such as partially visible or semitransparent objects cannot be viewed accurately and so their impact cannot be assessed. That's why it's a good idea to check out a few fully rendered frames as a part of the verification process. You may also discover other software constraints, such as lack of shadows or fog effects, that limit your assessment to simple motion verification and a medium level of rendering.

12

Advanced Animation Techniques

Once you have a good grasp of the fundamentals of 3-D model animation, you are ready to extend the tools for more effective power and control. More advanced capabilities (both 2-D and 3-D) are based on intimate knowledge of how the various computer graphics tools interact. These hardware and software tools include video input and output devices, video manipulation capabilities, and 2-D and 3-D animation software. You also need a clear concept of what these tools can and cannot achieve.

It should now be apparent that creating the illusion of movement is not always easily accomplished, despite all the notable advances of computer art and science. Ironically, when you finally have those dream tools in hand—the ones that allow greater control and freedom from repetitious drudgery—you will be confronted by new and equally daunting computer-related tasks. To explicate movement, for example, the animator must fill in the blanks, and this opens up a Pandora's box of often difficult questions. After you have finally mastered a particular new technique, you naturally want to extend it to other problems—and you may find out that, for some reason or other, you can't. For example, you decide to use texture repeatedly to achieve realistic mirror effects—only to realize that you cannot control which map comes first. You soon discover that no single method or technique can even begin to solve all of your animation problems.

You'll find that each animation project is unique: Whether the job is personal or professional, animations are strikingly different from one another. Each new job forces you to invent creative schemes to effect its realization. Some are "quick and dirty" situations—you need to create a simple animation in a hurry. Others may require the use of every tool and the extension of every capability in your arsenal of animation tools and expertise. Each job brings with it new challenges, new problems to solve, and, ultimately, more experience. At this point in the animation process, fewer rules apply, and you have a choice of many roads into the realm of creative thinking.

Time Management and the Bottom Line

Animation is inherently repetitious and time consuming. As a general rule, the higher the quality desired, the slower is the process. That's why you need to be able to gauge how long it takes to model, preview, and render an animated sequence. One way to get a handle on time expenditure is to keep a log of computer time: You need to be able to estimate how long a particular sequence takes to produce. The results of logging may surprise you. One key to efficiency is to zero in on rendering time. If a faster computer is (really) twice as fast, how will rendering time be affected? How would changing your processor affect modeling time? Waiting a second here and a second there can add up to a long modeling episode. When you take a close look at your actual time expenditures, you may find that the labor costs of modeling (that is, your time) may exceed the value of that new and faster computer.

Advanced Tricks

Thanks to digital compositing, one of the greatest assets of computer animation is that you can continue to add visual elements to a sequence without compromising quality. For example, say you create a painted 2-D image of a moody background and save it on the hard disk. Next you add some buildings with complex light sources. Now you add some moving characters—in 2-D or 3-D. All of these elements can be digitally added, adjusted, and smoothed; further effects can be incorporated later to give the sequence the desired finish. Finally, the whole scene can be rendered onto videotape with no overlay penalties in terms of image quality.

With experimentation, you can weave dramatic effects into a concrete animation. For example, after a background is rendered and saved on disk, you can go back and pull up the frames one by one and use them as environment maps to be reflected by a moving object. You can also incorporate live-action video images (see Figure 12–1).

If you call up a selected video frame from a VTR, the image can be used; it can be either captured digitally or "live" from a video camera. This image can then be applied as a background or used as the basis for a texture map. In fact, interweaving video imagery from live action with people (in street scenes or work scenes, for example) can be an effective visual technique. Still images, too, can be enlivened when slight touches of movement are added to convey a sense of action.

You can use this compositing technique on the desktop to create what would appear to be a massive 3-D model (similar to one generated by an expensive mainframe computer). If, for example, you wanted to visualize a giant array of hundreds of planets, you could render the one model several times at different camera distances and composite these images together. After you render the model at a far remove (from the viewer), you can use the rendered 2-D image as a background for the next round of rendering. For still more planets, repeat this process as many times as you wish. The resulting image is an impressive array of hundreds of planets.

Figure 12–1. *This frame of an animation incorporates painted characters over a computer generated background.*

Optimizing Animation Rendering Time

The total time it takes to render an animation on video or film is directly proportional to the time it takes to render a single image. However, several shortcuts can speed this process and thus lower production costs when a commercial project is undertaken. The following tricks of the trade are based on common sense and on a familiarity with the overall rendering process.

Perhaps the most commonly used technique to reduce animation rendering time is called shooting on twos (mentioned briefly in Chapter 7). The term and the technique are derived from a classical animation methodology in which each acetate cel is exposed to one frame of film. *Shooting on twos* means that, instead of one exposure, two exposures are made of each cel; thus only half the number of cels (and work) are required for a given sequence of animation.

Computer animators also can use the technique of shooting on twos. The motion appears somewhat staccato because fewer images are used to convey motion, but the reduced level of quality is not necessarily noticeable and may be quite acceptable (depending on your production objectives). Cartoons, for example, are shot on twos (or more). If your budget is small and production time is especially short, a rendering of 15 frames per second instead of 30 (for video) may suffice despite the reduced quality of motion.

As a rule, the less you have to render in 3-D, the shorter will be the overall rendering time of the animation project. In the example shown in Figure 12–2, only the moving features are animated; the background is static. Thus a game of ping-pong is animated in which only the ball is moving. You could speed the animation considerably if you did not elect to render the complete 3-D model of the scene. The following paragraphs describe how to do this.

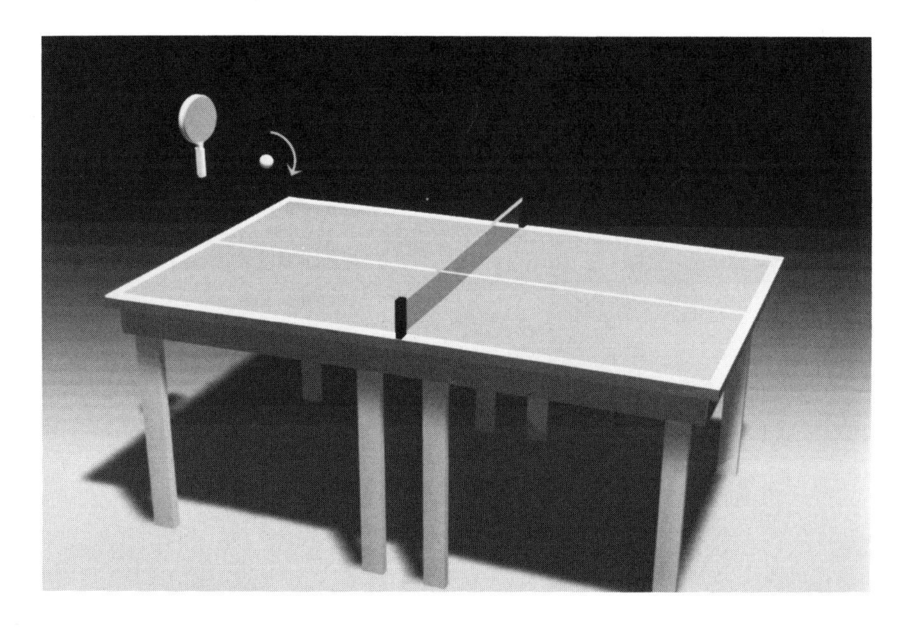

Figure 12–2. *This frame in an animation consists of a rendered 3-D image used as a 2-D (nonmoving) background and a 3-D paddle as a foreground. Using this 2-D and 3-D combination, the total rendering time is shortened (25 times faster) because only the small 3-D paddle is rendered, instead of the whole scene.*

First, you render the ping-pong table (without the ball) and the surrounding room in 3-D. You then save that image as a background for the action. Next, you clear the program and get ready to create a new model. The new model will consist of a white sphere placed on a background of the ping-pong table that you just created. Note at this point that when the ball is rendered over the image of the 3-D scene, you won't be able to tell that only the ball was newly rendered as a 3-D object. Next, you animate only the motion of the ball. As the ball approaches the viewer, you increase the scale (only the x- and y-axes are necessary) of the ball so that it appears to move closer to the viewer's POV.

Now you create a small, semitransparent, flat, black ellipse (or several concentric ones for smoother edges) and animate its motions to mimic the ball's shadow. Although you can't make the ball "go behind" the net, its motions can be fast enough that it's appearance at the net will go unnoticeable. (The addition of appropriate sound effects here will also enhance the illusion and can serve to mask a multitude of minor visual defects.) You'll find that by rendering an image that consists of a sphere and a set of elipses, you can work significantly faster than you would if you attempted to create a complete 3-D model of a room and a ping-pong table.

For even faster rendering time, you can animate the ball on a static background that is produced entirely in 2-D. In fact, you'll find that 3-D is not usually necessary in this simple kind of animation project. An image of the ball can be made with a paint program (or with the 3-D program). Then the 2-D animation program can move the 2-D image of the ball from one location to the next while changing its scale to invoke a feeling of depth. The background of the ball, which consists of a surrounding black square, can be designated transparent. What's particularly convenient about some 2-D animation programs is that this kind of animation can be run in real time at the same time that it is recorded on video. As you can see, knowing the extent of each program's capability can turn hours or days of work into minutes—often with little or no deterioration in terms of apparent quality.

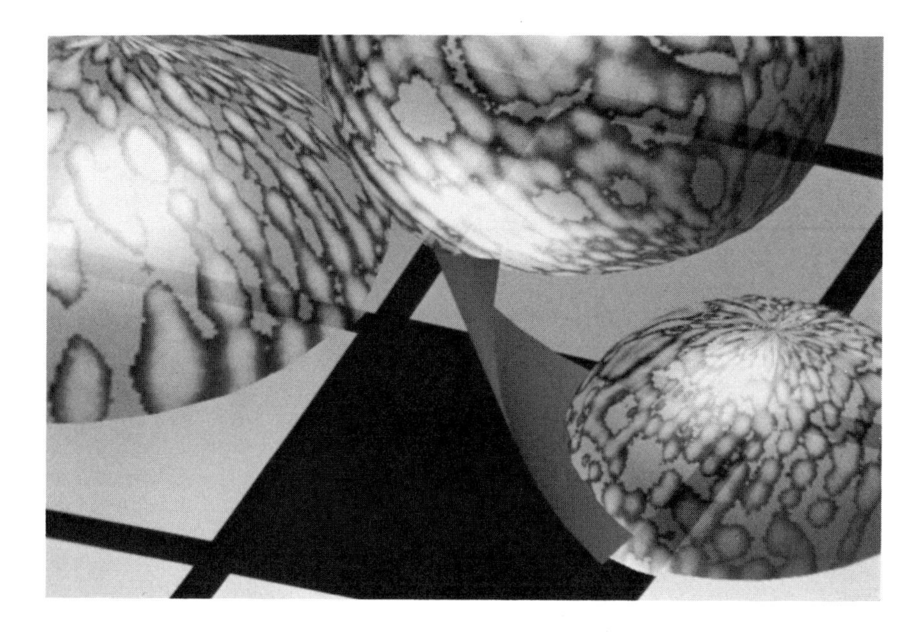

Figure 12–3. *Complex video production effects, such as this page-turning effect, can be created with precision on the computer.*

Special Effects

Some effects are not easily pigeonholed because they border on the land between video, 2-D, and 3-D (or they simply manipulate the image). Genuine 3-D effects are usually integrated with the animation. An excellent example of one of these hard-to-classify effects is the turning page shown in Figure 12–3. The following paragraph explains how it's done.

You start with the page, a rectangle that, in this case, will be texture mapped with a sequence taken from live video. Using 3-D techniques (as a function of the animation program), you can make it appear as if the page is being turned—it curls and then peels off. What really happens is that the rectangle is broken down into smaller 3-D polygons, which are then made to turn in a user-specified direction, to produce an effect that is much like the paper curling over to its other side. Because the movement occurs in true 3-D space, the rectangle possesses the appropriate highlights and reflections as determined by the position of the light source. A process such as this allows you to create the illusion of a 3-D effect and teases the viewer with what appears to be a live video image. Visual surprises like this can liven an otherwise dull video.

Your software will enable you to incorporate any number of advanced video effects into an animation sequence as long as the graphics board you are using supports video functions. Using 3-D animation techniques, you can use several tricks to take advantage of external video sources. For instance, you can set up a reflection map, which can then be used as an environment map. With this technique, a sphere, for example, can be made to reflect the frame-by-frame video image as though it were a chrome ball flying through the original video. Due to the "real" reflections, the ball appears to be integral to the live videotaped event (see Figure 12–4).

Figure 12–4. Live-action scenes can be automatically digitized and used as a background, and 3-D reflective objects can be incorporated in the final scene. The rendered sequence of images can then be placed automatically onto videotape.

Another important video interfacing capability (covered in more detail in the next chapter) is the ability to reserve specific regions of the computer generated image for later use by live video. The concept is straightforward. You first create an image—of a window, for example—and then assign it as black (or any other "flat" color). After the animation is completed, a video process called chroma keying assigns live video (from a VCR or camera) to that particular color in the animation. In this case, because we have selected flat blue, anything that is blue becomes live video. Keep in mind here that the chosen chroma key should be a color that does not otherwise appear in the computer generated image. For this reason black is seldom chosen—solid blue is a better and the more usual choice.

Many advanced enhancements can be incorporated that are generated outside of the animation program. Some animation systems allow you to render a frame, leave the program to run an effect routine, return to the animation process and place the resultant image on video tape (see Figure 12–5). An example of this kind of maneuver is the use of an image processing program which runs independently of other programs. Such a program is QFX (which runs on MS-DOS based computers) and provides image enhancement and manipulation capabilities. QFX allows the animator to adjust contrast, to blur and sharpen images and to add effects such as glows.

If your animation program does not allow external program manipulation, you can use other strategies to incorporate these kinds of effects. One way is to render your animated images to disk, then image process the enhancements as a group of separate images, and finally, via the external program, port the images to videotape.

Alternatively, you can use the processed images as individual backgrounds and overlay new 3-D images on them. For example, you

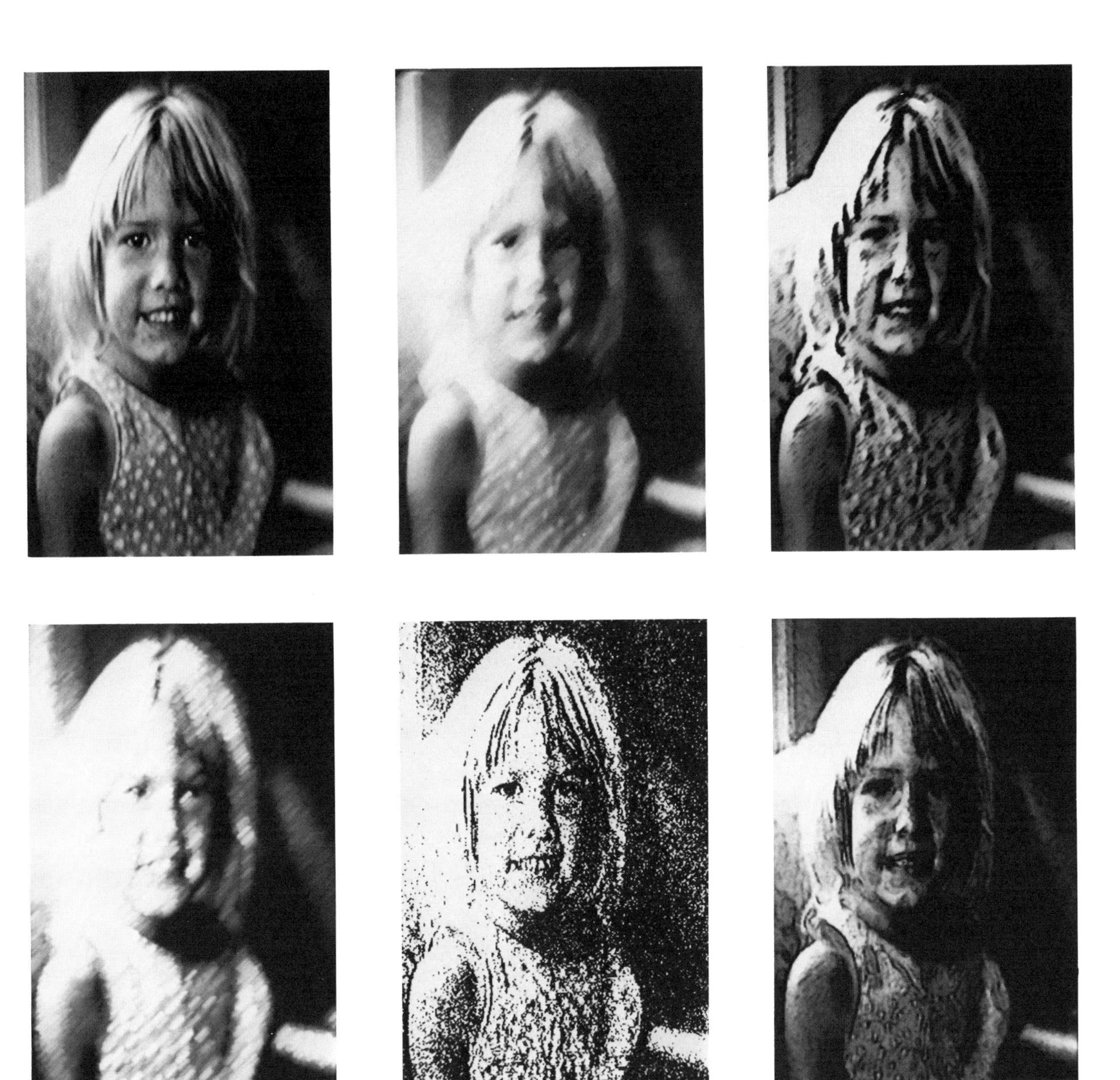

Figure 12–5. *Live video can be modified automatically frame by frame using image processing effects such as these from Imageware. The effects can range from sketchlike to surreal chromelike imagery.*

might elect to forge a sense of foreground and background using the same methodology that a camera uses. By adjusting a real camera's F-stop, you can make the background of a photograph appear out of focus. The animator can achieve the same kind of effect in three steps. First, render the background of the animation and save the image(s) to the hard disk. Next, process all the images so that they are blurred and appear out of focus. Finally, have the animation program use these images as sequential backgrounds for animated 2-D or 3-D characters. Figure 12–6 shows an example of this technique.

Figure 12–6. *Compositing images can add an extra dimension. The background is the same as the foreground but was first rendered, blurred, and saved on disk.*

Improving the Quality of Motion

A common complaint from the sensitive viewer of computer animation is that the motion has a kind of staccato effect; that is, the movements often seem unnaturally jumpy. It is one of the unwanted effects of rendering: In the process, a rendered image is frozen in time, providing no visual clue as to where the object is coming from, where it's going, and how fast it's moving. Particularly in fast-moving sequences, this frozen-in-time rendering appears inappropriate and unnatural. This unwanted effect is made even more noticeable when a wide-angle perspective is used.

Furthermore, if an object is approaching the viewer and made to pass by rapidly, it will seem to jump suddenly out of the distance to almost immediate proximity in perhaps as little as two frames. This unfortunate effect is called *strobing*. Our eyes naturally blur fast motion; in other words, we expect motion distortion. Motion pictures fulfill this expectation quite nicely. Single images are blurred in the filming process because the motion is going faster than the camera shutter.

If you were to carefully observe a rapidly moving object on your computer screen, you might notice that its movement appears to be staccato. This staccato effect is a kind of illusion: What you are really seeing is a split-second image of the moving object as it traverses the screen (and your view of it). It looks unnatural because we expect to see a quality of streaking when we observe rapidly moving objects.

The same kinds of aliasing problems confront the computer animator, except even more so. With computer animation, individual static images convey no sense of movement. In contrast, a fast moving object on a frame of film will appear blurred. For this reason, a film version of rapidly moving objects possesses a more appealing sense of motion than does a computer generated animation. Fortunately, using superior animation programs, you can improve the appearance of motion in a number of ways.

We can replicate this effect via computer with temporal aliasing. *Temporal aliasing* automatically creates streaks from objects that move quickly from frame to frame; this mimics the apparent distortion caused by fast movement. It's an effect you'll see used only in more sophisticated animations to boost visual quality. It's called temporal because the aliasing is done in reference to time.

One computationally expensive technique can significantly improve the overall quality of an animation. *Motion blurring*, as its name implies, adds a blurring effect: The software assesses what parts of an animation are moving and then adds a blurring effect based on the degree of movement in the most recent frames. The distances traversed in the preceding frames give the program an idea how fast an object is moving and, hence, how much blur to impose. The results are visually impressive; yet, the cost in terms of computational overhead is high. This process is often too expensive to make its use practical.

Figure 12–7. *The addition of motion trailing significantly aids the viewer's perception of rapidly moving objects. Without the motion blur effect shown here, the alternative would be to display two frames in 1/15th of a second; the object would appear only as flashes.*

Another technique, called *motion trailing*, produces an effect similar to motion blurring but requires less time (see Figure 12–7). Instead of adding blur, the program adds sequentially lighter versions of the moving objects on the basis of preceding images. For example, sequential frames of a rapidly moving UFO would display the object at first as solid and then would produce fainter and fainter versions of the preceding frames. This process, as you would expect, is demanding on the hard disk because it must store the last few frames (as determined by the animator) and repeatedly call them up for compositing with the newly rendered frame. This is another situation in which a fast hard disk can help improve overall rendering time.

A minor drawback to motion trailing is that it tends to reduce the brightness of a moving object. This may be a problem with video applications, where an important object might be blurred too much. Moreover, in some situations it won't work and looks worse than if it had not been implemented at all. For example, if you were to place the viewer close to a rapidly moving scene, you would get poor results. Otherwise, the effect can usually be controlled by varying the number of trails. When motion trailing is well implemented, the overall effect is a significant improvement in the visual conveyance of rapid movement.

Another type of temporal aliasing works by creating lighter and lighter versions of the object relative to its last appearance. Although this is not a perfect solution, it does reduce the jumpy effect—and does it with much less computational overhead.

Still another method can be used to reduce motion effects. The technique is based on the rendering of fields rather than frames (see Chapter 14). Rendering on video fields means that the image is rendered in half the resolution but at 60 times per second instead of 30 (for video). The result is greatly enhanced effects for fast-moving objects. Comparing a rapid panning sequence can be very convincing. If the pan is set up to traverse a cluster of buildings, for example, the sequence that has been rendered on fields is visually comfortable and believable, and the buildings, rendered normally, appear and disappear as the camera's view traverses the scene.

Character Animation

One of the most difficult challenges in both classical and computer animation is the conveyance of a "living" character's motion. An animated animal character, for instance, that is supposed to caricature a certain kind of person's walk (that of a bully, for example), is detailed by nuances that we usually take for granted as viewers. If controlling the movement of complex characters isn't difficult enough, the fine details and characteristics that convey lifelike motion can be equally difficult to articulate in words. Imagine, for instance, the simplest of characters—a robotlike creature walking. Even a robot's walk conveys emotion, depending on the gait assigned to it. Additional accessories, such as hats (which may or may not be connected to the head), signify characteristics of their own in relation to the robot. So how do you make these objects come to life?

Classical animators simply used the brute-force method: They create numerous keyframes describing as much of the movement as possible. Despite all our technology, computer animators often have to resort to the same method to get lifelike, believable animated characters.

The technique underlying all good character animation (whether the animator is classical or uses a computer) is called *squash and stretch*. This fundamental practice applies to relations between objects as well as to the basic components of the objects themselves. Simply stated, squashing and stretching allows the animator to exaggerate the effects of gravity and emotion. When a robotlike character is walking, for example, the motion not only should suggest an ambulatory direction but also should convey a sense of gravity. If this weighty quality is not explicit enough, even the most casual observer will notice that the robot is an unrealistic amateur production. When previewing an animation sequence, you need to ask yourself, "How is gravity affecting these objects?"

Squashing and stretching techniques rely on the distortion of rounded objects such as those made by splines. You can stretch a spline-made sphere, for example, by pulling on one of its (spline) control points. Because a spline-based sphere is a mathematical construct rather than an assembly of plates (such as those of a polygon), it is easily reshaped. If a character is defined so that a hierarchal path relates each object that comprises it, all the components can be made to change together as a properly controlled group.

Because of the detail required to make characters act realistically, you will find that numerous keyframes and repeated previewings are needed for feedback in order to verify the quality of a character's motion. Character animation is where the caliber of your software's previewing capability and your dexterity in accessing it becomes vital to your work.

Metamorphosis

Metamorphosis is a powerful technique that allows you to transform one object into a totally new one. Most often, this process is attempted only in 2-D because 3-D metamorphosis can get too complex for PC-based animation systems. A metamorphosis typically uses spline-based objects because they can be gradually and gracefully moved from one position to another based on the animation of their control points. For example, a simple spline-based sphere can be transformed into a star almost automatically by way of the metamorphosis function available in many software packages. Splines are used because the transformation can then be based on control points that determine the object's shape. When using such a feature, you are simply prompted to supply the number of frames to be included in the sequence.

Because of the unique nature of this function, it isn't used much, yet it offers the animator vast creative possibilities. For example, using metamorphosis, you can mimic a 3-D model and transform it into what appears to be a dissimilar 3-D image—providing the camera angle isn't changed too much to reveal the process. Perhaps of even more interest to the workaday animator, metamorphosis techniques can be used for transitions of all kinds, such as bridging from one logo or group of letters to another.

Creative animators have discovered ways to use the metamorphosis capability with texture mapping. Carefully mapping images of the faces of a house over a block can make the block approximate the appearance of a house. Then the house can metamorphose into an inflated balloonlike house, ready to explode, or its shape can be changed into any other totally novel object.

13 Video Basics

Video is the most popular viewing medium in the world. Yet, it is so complex that even technicians well grounded in electronics often don't fully understand the principles that make it work. Fortunately, viewers have little need to understand the technology that underlies video. However, if, like most computer animators, you find that you need to interface your equipment with the video world, you have no alternative but to become conversant with video technology, with the available equipment, and with certain video options.

Before exploring the process of recording onto video, this chapter looks at how video works and then focuses on the most critical interfacing factors—issues of video-compatibility.

What Is Video?

In a color graphics display system, the *RGB signals* sent from a desktop computer consist of red, green, and blue signals as well as the system's horizontal and vertical synchronization signals. These five signals (often sent via five individual cables in a pack) are designed to connect directly to an RGB monitor. You see an RGB image on your display screen, and so you might conclude that all you need to do is reroute these five video signals to a VCR/VTR for recording purposes. Unfortunately, it's not all that simple. Graphics display signals are not usually compatible with standard video gear, and so the video signals must be converted, or encoded, into a standardized video signal, called composite video.

Composite video, also known as either National Television Standards Committee (NTSC) or RS-170A, is the standard that has become the basis for ordinary television communications. Because composite video utilizes only one cable to communicate all video information (a clear advantage, logistically, over RGB) all RGB signals (including the sync signals) are combined in their transport through the cable. Thus, although composite video may be easier to manage, it has a significant drawback: its signal quality is inherently inferior to that of RGB. (This is why, in broadcasting, NTSC is jokingly referred to as "Never Twice the Same Color.") Color Plate 10(a) shows the quality of an RGB signal, and Color Plate 10(b) shows the quality of an NTSC signal.

There are two important components of a composite video signal: the black and white information, called *luminance*, or Y, and the color information, called *chrominance* (or chroma), or C. In order for a single wire to convey all video information, the luminance and chrominance (Y/C) signals are combined in a particular way based on certain subjective viewing qualities that were deemed desirable when the NTSC standard was adopted in 1953.

At that time, research available to the NTSC committee suggested that human vision is most sensitive to green, least sensitive to blue, and somewhat sensitive to red. They also concluded that human vision does not rely entirely on color information to resolve detail in a scene. Based on these assumptions, and desiring to standardize video signal transmission using current (1953) equipment capabilities and a limited bandwidth (amount of signal space used), the NSTC color format was biased and, of necessity, compromised. Yet, despite its notable limitations, the standard prevails to this day in the United States and in many other countries throughout the world.

In order for the graphics display information generated by your computer system to be compatible with the composite video standard, the RGB signals must include correct timing relative to the NTSC standard; that is, the signals must be synchronized, or in sync. Specifically, the horizontal frequency, or scan rate, must be calibrated at 15.7 kilohertz, and the vertical scan rate must be 30 hertz. One frame of video comprises 525 vertical lines, 41 of which are used for vertical delay. That leaves 484 lines that overlap slightly, providing about 350 viewable lines of resolution. Figure 13–1 shows what the signal standard looks like electronically and Figure 13–2 illustrates the timing of the NTSC standard.

To prevent flicker (which would otherwise be apparent), instead of refreshing the display with a new image 30 times a second, the odd and even horizontal lines (called scanlines) are displayed alternately. That's why each frame of NTSC video is composed of two fields: The odd and even fields are alternately displayed every 1/60 of a second, while every frame is displayed every 1/30 of a second. This odd–even display technique is called *interlacing* (this is described in Chapter 6).

Knowing Your Numbers. Technically, the 60-Hz rate is really 59.94 Hz. In 1953, when the NTSC introduced color to the monochrome standard and wanted to maintain compatibility with existing TV electronics, they discovered that the new color encoding format interfered slightly with the old standard. To eliminate this, the scanning rate was shifted a tad from 60 to 59.94 Hz.

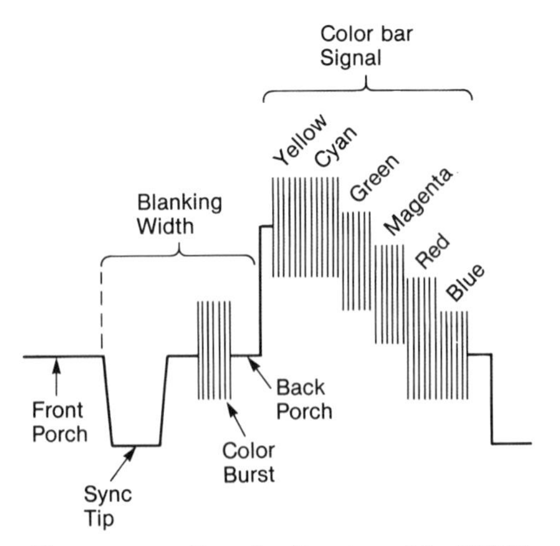

Figure 13–1. *Standardization of the NTSC signal is enforced by the FCC in the United States. The signal quality, as seen on a waveform monitor, should look like this approximation.*

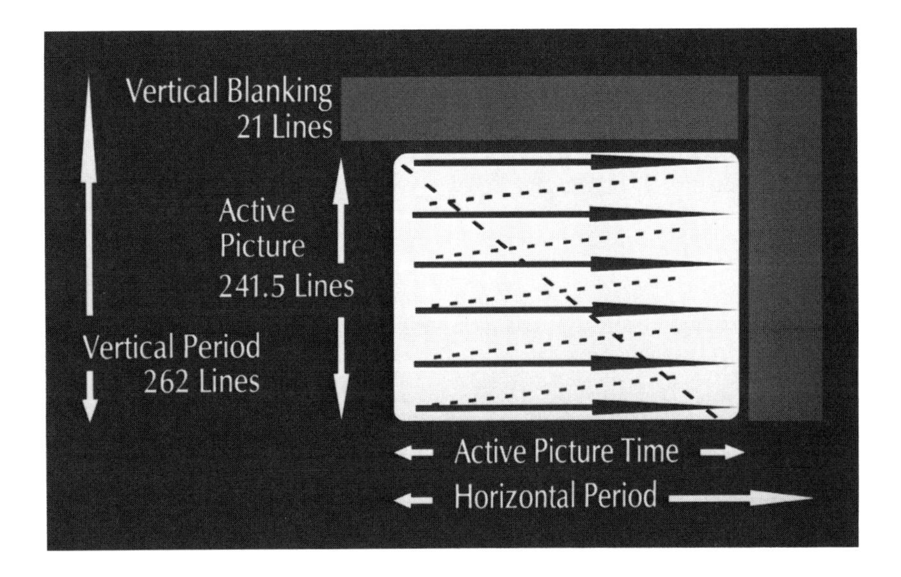

Figure 13–2. *The timing of each frame of NTSC video is complex, detailed, and predictable.*

Although European videotapes appear to be the same as ours on the outside, you will discover that, because of differences in video format, your VCR is unable to read or to record foreign videotape formats. In fact, regardless of the manufacturer, a VCR will find unreadable any format other than the one it was designed to accept.

So why are there different formats? The European video formats (theoretically) allow for improved color quality. Both the Phase Alternating Lines (PAL) and Sequential Colour à Memoire (SECAM) formats display the image 25 times a second instead of 30 and provide for 625 vertical scan lines instead of NTSC's 525. However, those accustomed to viewing NTSC often notice flickering when viewing European video—a distraction that is directly related to the lower scan rate. The PAL standard is used throughout most of Europe; France uses the SECAM format.

Foreign Video Compatibility. When animating a video for foreign distribution, producers have several options. They may choose to create the video using the most available technology (such as NTSC in the United States) and then transfer the animation later at a facility that specializes in format conversion. The problem with the conversion option is that there is an inevitable loss in quality, even with a high-quality master. Another option is to render the animation on film and later convert it to the required video formats. This approach bypasses a host of potential problems regarding video format compatibility and, if planned well, can save money in the long run. Another option is recording the animation on an optical disk in RGB mode. By using this method, when a video is required in any format, there will be no loss in its conversion.

Competing Video Formats

Because NTSC quality is so poor compared to the quality of raw RGB, (see Figure 13–3) alternative video formats are vying for the middle ground. One of these is Super-NTSC, also called S-VHS or S-video.

S-VHS improves upon the one-wire one-signal solution because it separates the chroma (color) information from the luminance (black and white) information. This format relies on a different multiwired cable and plug configuration for transport of the separated signals. The improvement in terms of apparent quality is dramatic. A video that is recorded and displayed on an S-VHS monitor is better, both in color and black and white, detail than the same animation generated and viewed using conventional NTSC equipment.

Another strategy employed to maintain high-quality signal transmission is *component video*. Introduced by the Sony Corporation, the component format completely separates the chrominance and luminance signals into separate wires, thereby providing higher bandwidth for both. The result is very clean colors and a high-definition black-and-white video signal. This format, although incompatible with typical network studio production, is equivalent to the best broadcast signal quality and so has become popular in high-quality computer animation studios.

Keep in mind, however, that regardless of the technical merit or the quality of the output generated, the primary reason these newer formats are so important to the animator is that they represent potential compatibility issues, particularly relative to the use of low-cost VTRs.

You will occasionally hear nontechnical terms used to describe composite video output. The terms "industrial" and "broadcast," for example, are often used to relate level of quality, and the terms are often confused. Broadcast video (sometimes called legal video) refers to a product that meets an exacting standard established by the FCC (which enforces adherence to the NTSC signal in all commercial broadcast transmissions). Video that deviates from the NTSC specs, such as a

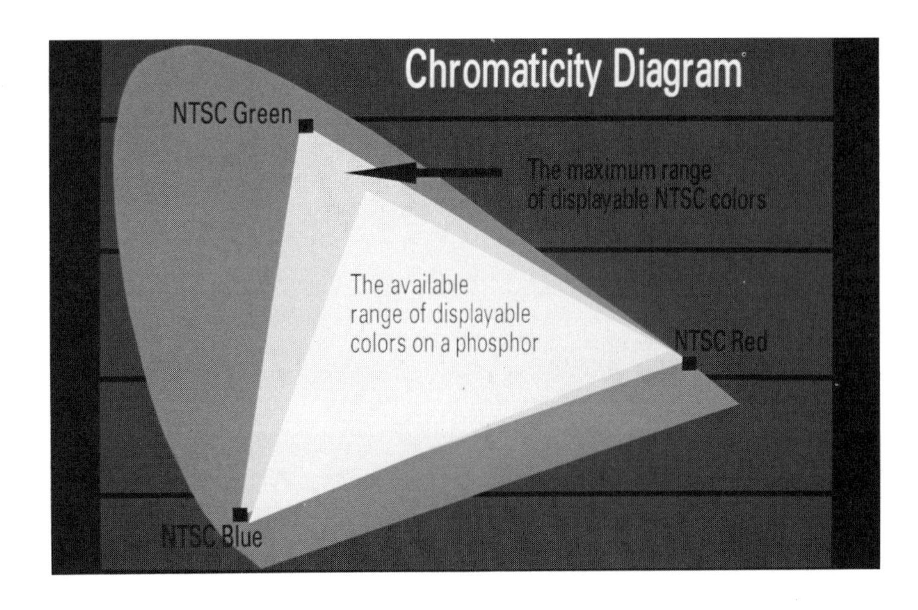

Figure 13–3. *The color range of NTSC video is limited. Furthermore, highly saturated colors are limited and become white.*

jittery chroma (color) signal, a delayed synchronization pulse, or occasional small lapses in voltage levels, may not be apparent to the viewer but cannot legally be broadcast.

Anything less than strict compliance to the NTSC standard is not considered broadcast quality and so is automatically downgraded to industrial-grade video. What does all this mean to the desktop animator? If your animated video sequence is not intended for broadcast on major networks, producing to NTSC specifications is not usually worth the considerable additional expense.

Caution with Compatibility. If your final medium is video, be sure to view your computer graphics images on a composite monitor to verify color quality and definition. Also be sure that the tint, color, and contrast on the composite monitor is set and maintained correctly throughout the project. The display of color bars facilitates that process.

Encoders

Devices called *encoders* are used to convert the separate RGB signals into a composite NTSC video compatible with standard video gear. Encoders are either integrated on your computer's graphics display circuit board or housed in a stand-alone box. One index of the variation in capability built into the electronics of stand-alone converters is cost. Prices range from $400 to $20,000, depending on output quality (for example, broadcast level, filters advanced electronics, etc.) and on the degree of manual control afforded the operator for the manipulation of various signal attributes.

For an encoder to work properly, RGB input must be electronically accurate. That is, the original RGB signals must arrive at the correct horizontal and vertical scanning rate (interlaced, of course) as well as at the proper voltage levels. The scanning rate of most graphics display units is usually controllable by your animation software (or by some other program to effect proper initialization of the encoding process.)

Low-cost encoders may be acceptable for industrial-grade video but will most likely be unable to generate broadcast quality. If your objective is to produce broadcast-level quality, sync signal control (such as delay and chroma offset) is a highly desirable feature. The better encoders produce superior NTSC signal quality by using electronic filters to remove unwanted frequencies. They also address the problem of chroma crawl, which is perhaps the most obvious of NTSC artifacts.

Chroma crawl is an insidious problem inherent to NTSC video. If you were to display a typical set of color bars, you would notice a distorted area between distinct colors. Lines appear to crawl along the color border. The artifact is not uncommon: For example, you might see a news broadcaster who accidentally wears an incompatibly colored tie or other colored clothing, which appears to self-animate due to this effect. A good encoder can reduce not only chroma crawl problems but also other artifacts, such as the colored moire effects that result from tightly spaced black-and-white lines.

All encoders, even the low-cost units, should have a feature called loopthrough. A device with *loopthrough* capability allows you to in effect bypass the encoder's circuitry so that both processed (encoded) and unprocessed output is available to other devices (see Figure 13–4). Thus, you have the ability to connect and daisy-chain other devices. This feature allows for flexibility in configuration: You might, for example, wish to connect several VTRs or monitors to your system.

With or without loopthrough, an encoder should provide for a simple electrical process called *termination*. This option allows you to switch on a 75-ohm resistor, indicating to the signal source that the wire has ended its transmission. In other words, the last unit (monitor or VCR encoder, for example) deployed in a chain of connected components should have the termination signal switched on (terminated). All other devices in the chain should have their termination signals switched off. If termination is not properly signaled on all pieces of equipment in the chain, the signal quality of any or all of the devices will be seriously impaired.

Some encoders provide additional video outputs (such as S-VHS) that use different types of plug and socket configurations. In fact, some VCR manufacturers use different types of S-VHS connectors, so an adaptor may be necessary.

Some encoders include their counterpart, a decoder (this was discussed in detail in Chapter 5 on graphics display units). A decoder accepts NTSC input and separates the red, green, and blue components of the video signal. A decoder is typically used for video capture from a camera, VTR, or other device. The separated signals can then be digitized in the computer for image processing or other applications. Keep in mind, however, the computer maxim, "garbage in, garbage out." At best, the visual quality is only as good as the original NTSC signal.

Another kind of encoder is called a scan converter. As the name implies, the *scan converter* converts one scan rate (usually NTSC video)

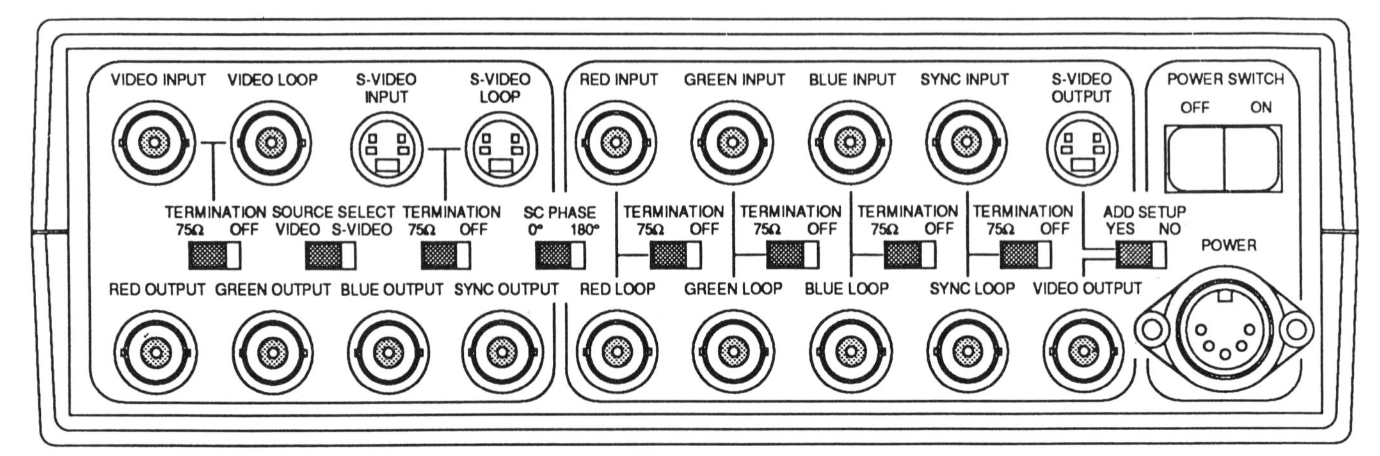

Figure 13–4. *The back of Truevision's Vid I/O conversion unit illustrates that it provides loopthrough, termination, and different choices for input and output.*

to another. Typically, scan converters are used to convert the images generated by high-resolution computer display screens, such as 1024 × 768 pixels (non-interlaced), so that the output is video recordable. These encoders can also manage input whose pixels and lines are substandard, that is, below the resolution of video. However, manufacturers of scan converters handle these kinds of problems in different ways, and the solutions can be expensive—sometimes over $10,000. Good units offer additional features that serve to reduce flicker, allow for adjustment of the video size and aspect ratio, provide for advanced color filtering, and enable video or computer graphics overlay.

Keeping in Sync: Genlocking

Synchronization is absolutely essential to video compatibility. All devices in a video system must be in sync. For example, the sync signal is used to inform video equipment when a video scan will begin and end. If there were no sync, devices (such as a VCR/VTR, camera, or graphics board) would display video images without reference to the pace and display of supporting equipment. The resulting visual cacophony would produce utterly useless output. Indeed, even minor variations from sync can cause problems. If the sync timing is only slightly off, many processes can be adversely affected and the resultant images may appear distorted or worse.

Cables and Video. Delays caused by different cable lengths can cause video problems. Different lengths of cable cause differences in the rate of delivery of video signal pulses (by a few nanoseconds) and can thus adversely affect video timing. For example, if the red wire from an RGB monitor is longer than the green and blue cables, the red part of the image will be slightly offset. Video sync is so sensitive that even time spans of nanoseconds are too long. Color problems will be noticed first. The color of composite video, determined by the phase of the chroma signal (also known as the subcarrier), runs at 3.58 megahertz. A delay of only a nanosecond or two due to a mismatched cable will cause a color shift. For this reason, you should always use cables of the same length to connect video equipment.

To control synchronization and forestall the myriad problems engendered by subtle timing lapses, professional studios establish "house sync" via a special video sync generator (or another source such as the VTR). House sync is simply the master sync to which all units are timed.

Professional-grade sync generators are used to provide stable oscillation and other timing-related capabilities. Some of these devices enable controllable delays and other adjustments that are used to

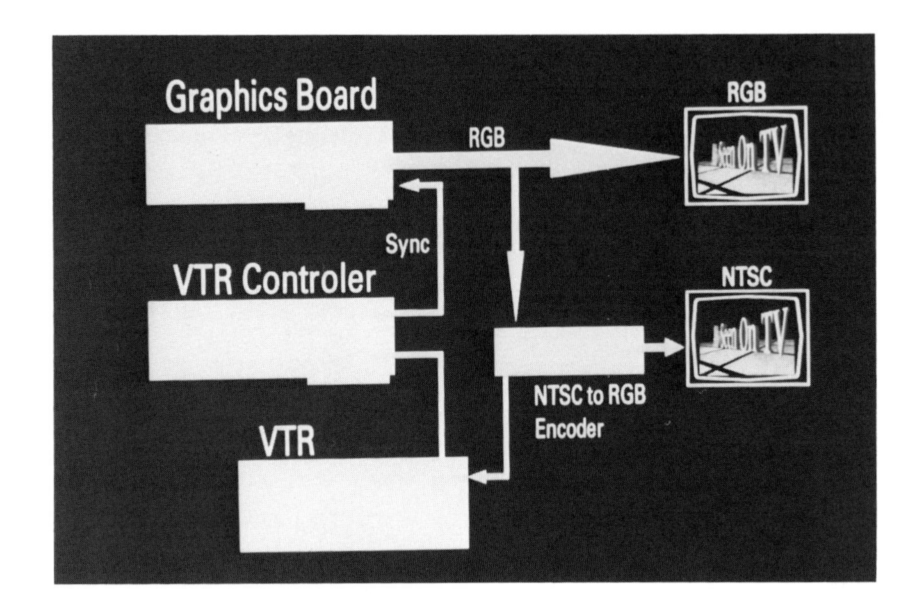

Figure 13–5. *A sycronization signal can be obtained from the VCR controller so that all other video components have the correct timing.*

fine-tune the sync output to compensate for errors cause by cables or additional equipment. For example, the phase of the chroma which controls the hue, can be offset to compensate for long cables.

A low-cost and reliable solution to potential sync problems is to use an immediately available external sync signal source, such as an RGB-to-NTSC encoder. In this case, the encoder acts as the master source and controls the sync of the VTR and the (genlocking) video board. The video board's ability to accept this sync information, called genlocking (this is discussed in Chapter 5), is a feature essential to the creation of high-quality video. Without genlocking, the video board free-runs its RGB signal timing without reference to other support units that need to be coordinated—that is, in sync—with the video (see Figure 13–5). Improperly synchronized video recordings are typically identified by their jumpy video image. In some cases, the image may appear satisfactory (even on an inexpensive VTR), and yet, because of unstable sync, the recorded images may be impossible to duplicate.

Overscanning/Underscanning and Display Size

The common, yet ambiguous terms underscanning and overscanning, are confused even by animation professionals. Both words refer to the area of the viewable video screen. Ordinary home TVs utilize less than the full picture transmitted to them, and so some of the picture extends beyond the viewable borders of the picture tube. In other words, the image is cropped, or *overscanned* (see Figure 13–6a).

When creating an animation using a good RGB monitor, you can see the whole image on the screen, including the black border surrounding the image. There is no doubt in your mind where the image begins and ends. This image, which includes the border surround, is said to be *underscanned* (see Figure 13–6b). This is where overscan discrepancies

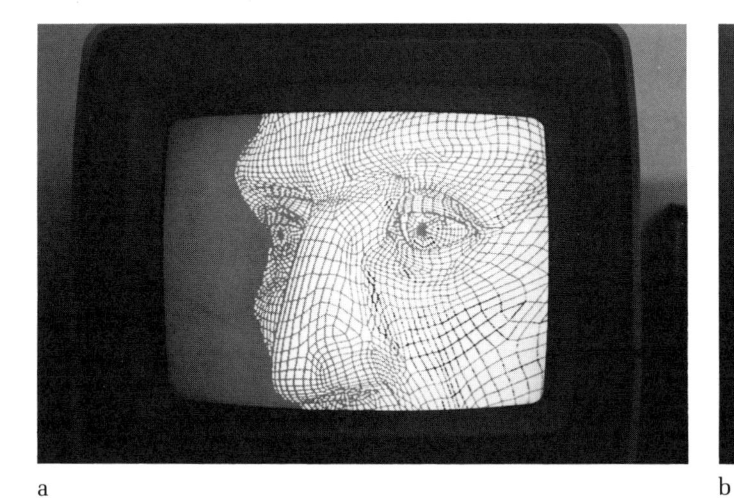

a

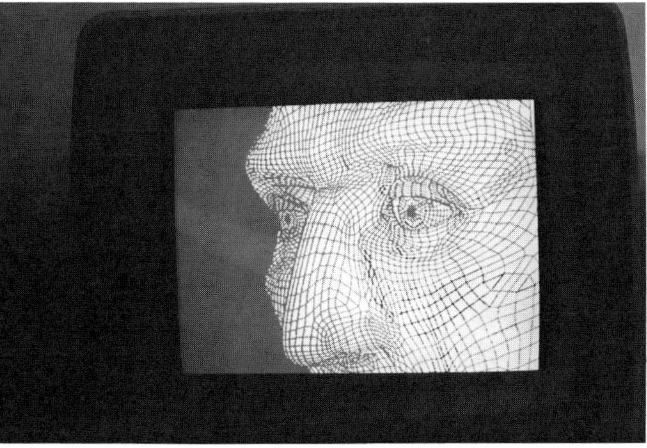

b

Figure 13–6. Some monitors provide a means for switching between video (b) underscan and (a) overscan. Note the borders of the two images.

tend to cause problems. When the same image is viewed on a composite monitor—the typical viewing medium for most animations—parts of the image are likely to be cropped, that is, overscanned. To compensate for this, some monitors (both RGB and composite) incorporate over-or-underscanning switches, which allow the video screen to either scan all the lines of the image or overshoot them.

Because of monitor-to-monitor differences, the video industry has developed a standard called the safe titling area (STA) (see Figure 13–7). Anything within the *safe titling area* is definitely viewable. However, image information outside this area may or may not be viewable depending on the degree of a monitor's overscanning. Thus, when creating text, you want to be sure that it falls within the STA boundaries.

Animating for U.S. and European Video. If you are rendering for both the U.S. and European broadcast markets, plan on rendering the additional scan lines for use with the European video format. Otherwise, the converted NTSC image will appear smaller. Also, there are losses in quality when converting from NTSC to PAL or SECAM. For optimal quality, you may want to save the animation digitally so that it can later be recorded on separate types of VTRs.

Waveform Monitors and Other Support Equipment

Questions about quality inevitably arise because of the importance of satisfying video standards: How can you tell if your video quality is meeting spec? For broadcast applications, where this is an essential question, two oscilloscopelike devices, a waveform monitor and a vectorscope, are used to help you come up with a suitable answer.

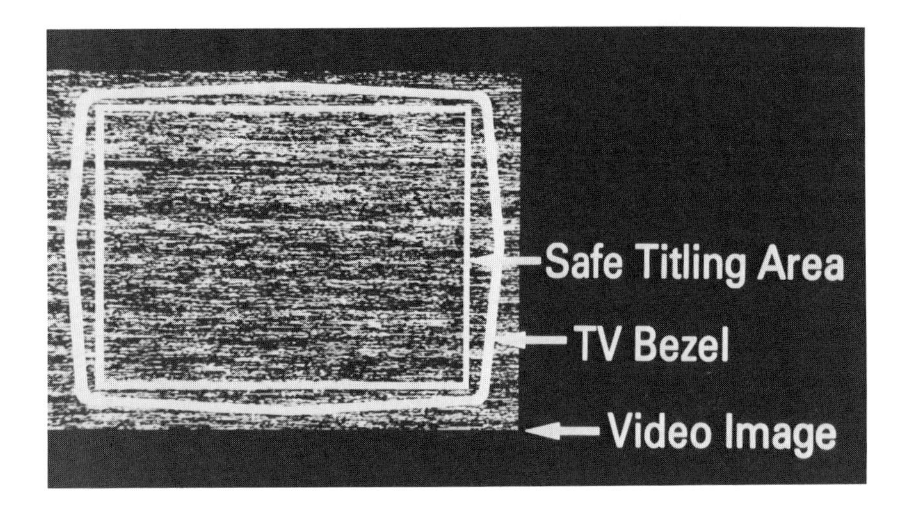

Figure 13–7. *The safe titling area is safe because it accounts for the image cutoff typical of most home television sets. Important information, such as text, should be placed within this area.*

The *waveform monitor* is similar in appearance to an oscilloscope, but it is particularly designed to monitor specific signals (such as that generated to satisfy NTSC or some other video format). A depiction of the video signal is displayed on screen: The lower rectangular sync pulse is seen along with the remaining horizontal scan lines (see Figure 13–1). The waveform monitor's controls allow you to zoom into specific parts of the signal—such as the *chroma burst*, which displays the video signal's color attributes. Given the necessary information regarding signal offset or chroma jitter, for example, you can then make necessary adjustments, such as repairing or changing cables or replacing faulty equipment.

Likewise, a signal's voltage levels can be adjusted so that the signal meets FCC-specified NTSC levels. You should know that even when your video equipment has not been moved and the temperature is stable, voltage levels tend to drift with time and so require occasional adjustment.

In a typical setup, next to the waveform monitor you will find a *vectorscope*. Round in shape like a radar screen, the vectorscope's electronics produce markings on the screen for R, G, and B, that is, for the red, green, and blue positions. When color bars are displayed, the vectors displayed on the monitor should have their "elbows" positioned within each of the marked positions. Adjusting the hue on the encoder rotates the elbows to the correct position. Of course, there are specially configured waveform and vector monitors designed for other video formats (component, SVHS, PAL, and so on).

Time Base Correctors

If you require professional (or reliable) results when videotaping, you will need a *time base corrector* (TBC). The TBC adjusts all the timing offsets from the mechanics of videotape recording, such as the stretching of the tape and fluctuations of the motors in the VTR. Your productions may look fine on the monitor, but you may discover later that other video equipment, such as editors, can't accept the varying timings generated by the VTR and so will display the attempted transitions with a broken-up image.

Until recently, TBCs were large and prohibitively expensive devices found only in post-production facilities where final editing was done. Today's units, however, are considerably smaller and less expensive, and they are essential in the production of high-quality video animation. As mentioned before, the exacting specifications of the NTSC signal must be adhered to by all related video equipment. This can seem to be an almost impossible task, given the proliferation of numerous cables and varying equipment setups.

The TBC's job is to ensure absolutely correct timing even though the output from the video source varies or is slightly unstable. All sorts of potential timing snags can be sidestepped. For instance, you may be able to record and play back an animation on a VCR with a slightly offset sync, yet when it comes time to duplicate the animation onto another VCR, you'll run into problems such as color changes or jumpiness. It can get even worse: The video might not be playable at all on another VCR. You'll find that although video recording is possible without a TBC, quality work requires one.

The TBC takes the "unclean" video signal and electronically rebuilds it so that the timing and amplitude of the output signal is absolutely perfect. It strips the sync off the original video signal and then reconstructs it based on another sync source (often self-generated). Only then can you be sure that the VTR will dependably and accurately record the computer generated image. Finally, if you have produced an animation for which industrial quality video was deemed good enough and so you did not bother to closely adhere to NTSC standards, a good TBC can sometimes be used to "fix up" the level of quality for broadcast applications.

Capturing Images with Video Capture Boards

Some graphics display boards offer *video capture* capability. This feature allows you to capture live video images from a VCR or video camera and then save them in a digital format. A frame-grabbing video digitizer is a useful addition to an animator's set of tools. Captured images can be used to turn ordinary polygonal blocks, for example, into realistic-looking marble (a process called texture mapping). Video-captured images can also be used for background scenes or for use as temporary guides in the modeling process.

Captured images can originate from many sources, including slides and photographs as well as live video. Typically, however, there are two main sources for most captured video: images are grabbed from either video cameras or VCRs.

Capture via Video Cameras. The most common method used to capture graphic images is via video camera. You will find that video cameras vary considerably in price, features, and quality. For computer graphics purposes, a simple camera without expensive features is sufficient. Although output is typically below the quality of RGB, some cameras provide additional S-VHS outputs that provide a good balance between convenience and quality.

Useful features in a video camera include color balancing capability (for incandescent or sunlight sources) and a manual exposure override feature (for altering apparent brightness). Also, some zoom lenses include macro capability for close-up viewing, which can be useful for capturing slides or small objects. However, you may discover that because these lenses were not primarily designed for close-ups, image quality tends to be poor and edges of the slide may be out of focus.

The usual measure of camera quality is resolution, which can be misleading. Both low-resolution and high-resolution cameras have potential output problems. A low-resolution camera (less than 300 lines) produces, at best, low-resolution images on the video screen. The lower the resolution, the greater is the likelihood of unwanted artifacts.

A high-resolution camera provides superior output (as its price will often reflect), but high-resolution cameras present potential problems too. When you connect a high-resolution camera to a high-resolution display system, you may find that the devices interefere internally with one another. That is, the lines of resolution may "beat" with each other, creating annoying Moire patterns due to the difference in resolution of the camera and the digitizing board. In either case—high-resolution or low-resolution—it's best to test a camera with your system to be sure you are not going to run into peculiar video inconsistencies.

The numbers that describe the resolution may be yet another source of bewilderment. Sometimes a camera is rated at 480 lines because it delivers 480 electronic lines of video. But the monitor's tube may really have the capacity of resolving only 330 lines of detail.

You should know that the resolution of a camera is measured at its best location, usually in the center of the screen. Thus a camera may be rated at 330 lines, but if the resolution at the edges of the image is only 200 lines, the overall effect may be unwelcome, to say the least. If the quality of the video-captured image is important, purchasing a more expensive camera and lens is essential. Many cameras detect the video image using standard Vidicon or similar electro-optical devices. However, newer charged coupled device (CCD) cameras offer several advantages of particular interest to those involved in computer graphics. These new digital image converters enable better image geometry and generate straighter vertical and horizontal lines because the resultant image is produced on a flat, square sensor. They also last longer before screen burnout, and so they reduce an unwanted effect called dragging. *Dragging* is observed when a bright object, such as a spotlight, is still viewable—dragged from its original position—even though it is no longer in the scene. Yet for many computer graphics purposes, if the camera is to be used for capturing still images, expensive tube-type cameras can be purchased second-hand at low cost, and these can produce excellent results.

CCD-based cameras offer other advantages. They don't create burn-in from intense overloads, and they deliver tighter color resolution, as shown by reduced color shift (shown in different colors). One of the few negative characteristics of CCD cameras, however, is that they tend to generate vertical line artifacts from very bright light sources. This can be controlled by changing the overall intensity through the iris of the lens.

Another potential drawback to the use of digital CCD cameras is that they encompass a somewhat smaller dynamic spectrum: The range between the blackest black and the whitest white is less than that produced by a tube-based camera. For scenes with relatively low contrast, however, such as for use in capturing pictures, their output is usually satisfactory.

In any case, for video capturing, you should use a light source bright enough to require the camera to use its smallest F-stop (if possible). This will maximize the overall capability of the lens by expanding the depth of field, thus making the most of the focusing ability of your lens.

The quality of the light source is also an important factor to consider. For best results, you should use a uniform, pure white light source (such as a dual incandescent and fluorescent light). Incandescent lights alone are typically yellow-reddish, whereas fluorescent lights alone tend to be greenish. Used together, they make an acceptable white light for illuminating slides or prints.

In terms of image quality, it's best to start with a video signal stored in RGB format (before it is converted into composite). This way the source information is in the same storage format used by other parts of the graphics system.

Fortunately, many newer cameras intended for use with computer graphics equipment provide RGB output. You may, however, be required to use adapters that will reconfigure the four RCA connectors to the particular socket arrangement specified by the graphics board or host. Note that you should always use good, well-secured cables: Loose connections or poor cables tend to introduce unwanted noise and strange patterns onto your otherwise clean video image.

Typically, a video camera internally creates an RGB signal and then converts it to the more popular composite format. So what about reversing the direction of the conversion? Don't make the mistake of trying to convert a camera's composite signal back to RGB. You will not be able to increase quality as you might think: Once the signal has been degraded to composite format there is little hope of reversing the operation and regaining RGB quality.

Capturing Images via VTR. The VTR is another source of visual imagery, especially for backgrounds and texture mapping. Images can be taken from a VTR either while it's in freeze-frame mode or while the video is running. If a fast-moving image is captured while the videotape is running, it is possible (and likely) that you'll capture two fields from different frames, creating two incongruous images. That's why it's best to try to find stable images in free-running videos for use as image sources.

You should be aware that many VTRs, by format definition, have poor resolution: Industrial quality 3/4-inch video formats produce about 330 lines of resolution, and VHS produces even less, generating only 240 lines. This means that even though the resolution appears acceptable when the tape is running at normal speed, a frame is likely to appear noticeably degraded when it is isolated.

When a VTR has been set in freeze-frame mode to capture a specific frame, VTR synchronization can be troublesome. The video capture capability of your graphics board may be unable to lock onto such a deviation from normal sync. In this situation, the best solution is to start with a stable source of sync sent to both the VTR and the graphics board. If possible, use a common sync throughout the system, including the video camera.

Recording to Video

14

An animation can be viewed on video, film, or the computer itself. This chapter continues to examine the most common, and yet perhaps the most difficult, recording medium, video.

Video is clearly the most widely used and most popular medium for rendering and viewing an animation. This should not surprise you, considering the size of the industry and the ubiquity of the medium. Despite its utility and flexibility, however, video is not conducive to the placement of single images of animation on tape. For this reason, you will probably need to become well acquainted with the technology and operations necessary to produce a video recording. Knowing how the medium works and how to manipulate the equipment can make the difference between a professional quality product and what, to even the most casual observer, is likely to be seen as an amateur's handiwork.

You should remember that video gear is expensive. In fact, the cost of video support equipment may exceed the cost of your desktop computer animation system—just one more reason to evaluate your animation needs carefully in terms of both tools and process.

Your Choices

There are two ways to record computer generated animation on video: You can record live or single-frame. Each method has its advantages and drawbacks, and the methods differ considerably in regard to overall process and equipment.

Recording an animation live from the computer screen is the easiest and least expensive way to go. The level of resolution and color quality, however, is limited, and so this method is best suited to noncommercial animations where viewers will not expect top-notch video quality.

The other method is single-frame animation. Because most computers can't process or display high-quality 3-D animation in real time, computer animation is usually created one frame at a time and placed onto video tape. In order for single-frame animation to proceed successfully, you need video interfacing equipment, a high-quality video tape recorder (VTR or VCR), and detailed knowledge about how properly to connect the computer and video support equipment. Single-frame animation's payoff is first-rate quality, the kind of output you're used to seeing in the best TV work. The following paragraphs take a look at these two recording methods in a little more detail.

Real-Time Video Recording. Although recording animation as it is displayed from the computer may seem the most obvious way to get an animation onto video, there are several technical hurdles to be overcome. First, the signals from the computer must meet the video-compatibility requirements of the videotape recorder. Assuming complete NTSC compatibility, the computer must be able to display the animation in real time, that is, at the same rate it is to be viewed. This built-in limitation creates a bottleneck in the animation process: Desktop computers are simply not yet fast enough to calculate or display complex high-quality 3-D animation in real time. Even if your computer system could display frames at close to 1/30th of a second (to accommodate video's usual rate of 30 frames per second), your animation would still be unrecordable on a VTR without special and prohibitively expensive equipment.

Fortunately for many applications, acceptable compromises can allow you to work with direct computer-to-video recording. The compromises are typically reduced resolution, limited colors, no anti-aliasing, or the animation of small regions or parts. These compromises can be very acceptable in certain situations (situations that don't require full resolution or full-color animation). For example, when animating scientific data, or when generating architectural fly-throughs, business graphics, or accident simulations, you can often convey the necessary information via the moving graphics without the need for top-notch broadcast-level quality.

You can use a number of simple techniques and tricks to produce this level of animation quickly. First, you need a board or display system that is video-recordable—that is, standard NTSC. Next, in order to record the animation directly, you'll need either a VTR whose sync is controllable by an external source (in this case the video from graphics board) or a VTR that provides sync to drive the input of the graphics display unit. Whichever you choose, sync conformity must be identical for the board and the VTR. Once this is assured, animation quality will be acceptable for viewing and will be duplicatable on other VTRs.

You will often want to insert frames, add to the animation, or edit the video for titles or effects. These functions are accomplished via single-frame animation. However, you should be aware that effects must be well thought out beforehand so that redundancy is avoided and quality is maintained.

Single-Frame Video Recording. Assembling single images on videotape is a complex operation and is thus potentially troublesome. You need a good understanding of the process to ensure a good finished product. However, once a system has been properly set up, you will find that you can automate the animation in a trouble-free operation. This allows you to focus on the more creative aspects of your work while your equipment cranks out the frames.

Consider the VTR. Most videotape recorders are designed to run continuously; that is, they are not designed to stop and start on demand for every frame of video. For this reason, for you to record a single frame, the VTR must be put through an arduous routine to get the frame recorded on video in the right place at the right time.

Here's how it's done. The tape is run backward a small, measured amount, stopped, and then run forward in order to reach the correct recording speed (see Figure 14–1). This little mechanical procedure, called *preroll,* is essential to the correct timing of a video recording. At exactly the right moment, the desired frame is recorded onto the video tape. A high-quality videotape recorder is used in this process. Unlike home VCRs, a higher quality VTR, can be directed to a particular location on a tape and there record a single frame of video. This feature is referred to as *single-frame accuracy*.

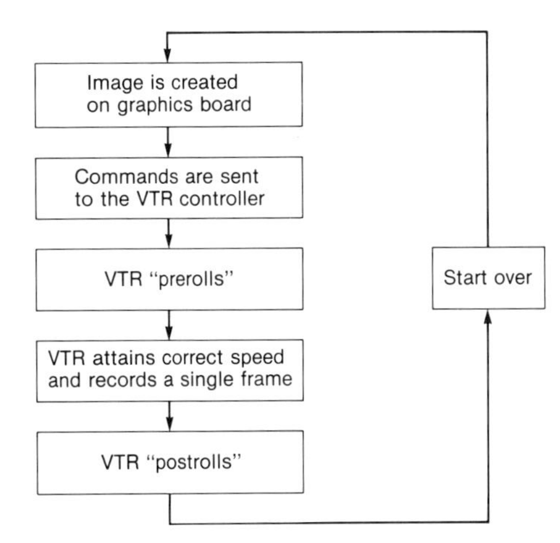

Figure 14–1. *Preroll (and postroll) is indigenous to mechanical tape recorders, ensuring that the tape physically records and plays at a constant rate.*

The VTR usually has a range of other features that facilitate the single-frame animation process and ensure a high level of quality. It usually accepts different types of sync inputs and outputs and provides at least two audio channels. Connected to an editing controller, a competent VTR can be used during editing and can also be remotely controlled (by the animator or some other remote device).

Another device essential for the computer animator is the VTR controller. This component is crucial to the operation of single-frame animation; it is the device that interprets animation software instructions into video instructions. The controller tells the VTR when and how to record a frame: It turns the VTR on and off, and correctly positions the tape for the recording of each individual frame. The controller supplies the electronics interface between the computer and the VTR and sits in the middle as a kind of mediator: The animation program communicates instructions to the controller, which in turn sends the appropriate signals to the VTR for correct operation of the recording. Controllers are manufactured either as a stand-alone unit or as a circuit board that plugs into your (host) computer.

In order for the VTR and the controller to know the exact location of any frame of animation on the videotape, a readable code is superimposed on the videotape. This *time code* is essential to video control and editing. The following sections look at these devices in operation.

VCR Controllers

Regardless of the physical interface, a VCR controller employs one of two basic strategies, depending on the type of VCR to which it is connected. Most professional grade VCRs are engineered so that they can be controlled externally, that is, remotely. Communications from a remote device such as an editor (or some other piece of equipment) to the VCR are managed via a communications port that operates either serially (RS-422) or parallel. Note that this particular kind of serial port is *not* configured as a typical RS-232 computer serial port and so is not compatible with a standard RS-232 interface. However, the latest computer-oriented VCRs are now available that provide RS-232 compatibility for single-frame recording. A typical system configuration is shown in Figure 14–2.

There is no functional advantage to either method as far as the animator is concerned, but you should make sure that the controller (board or unit) is compatible with your VTR's communications setup. For parallel operation, specific cables are required to connect to a given VTR. Generally, more sophisticated professional (and usually more expensive) VTRs operate via a serial RS-422 port and so are not cable-specific.

When selecting a VCR controller, you should carefully evaluate the features. Here are a few of the more important ones that it should provide:

- internal sync generator (NTSC or PAL)
- accepts external sync
- interfaces with a variety of graphics devices
- good documentation for installation
- diagnostic capabilities
- a time code reader/generator
- interfacing with a variety of VTRs
- all necessary cables for interfacing

Low-Cost Video Previewing Solutions. Previewing an animation is essential in the management of high-quality animation production. Yet, you don't have to preview the animation on a high-quality, and thus expensive, VCR. You will find that it's a good idea to use a cheap and expendable recorder when creating motion tests and performing other preliminary animation activities. This can save you capital expenses while providing excellent quality. For the final animation, when quality counts, you can rent a high-quality recorder. Here is an example.

A low-cost solution for motion testing relies on a VTR designed for low-cost still-frame recording. JVC makes a series of VHS VTRs intended for continuous operation for capturing frames without time-consuming preroll. These relatively low-cost VHS recorders (under $2000) can be operated via a serial port, and they can be used for motion testing and verification before an expensive recorder is used. The more expensive recorder can be rented exclusively for rendering high-quality animation (or for recording it later after rendering the animation sequence on the hard disk).

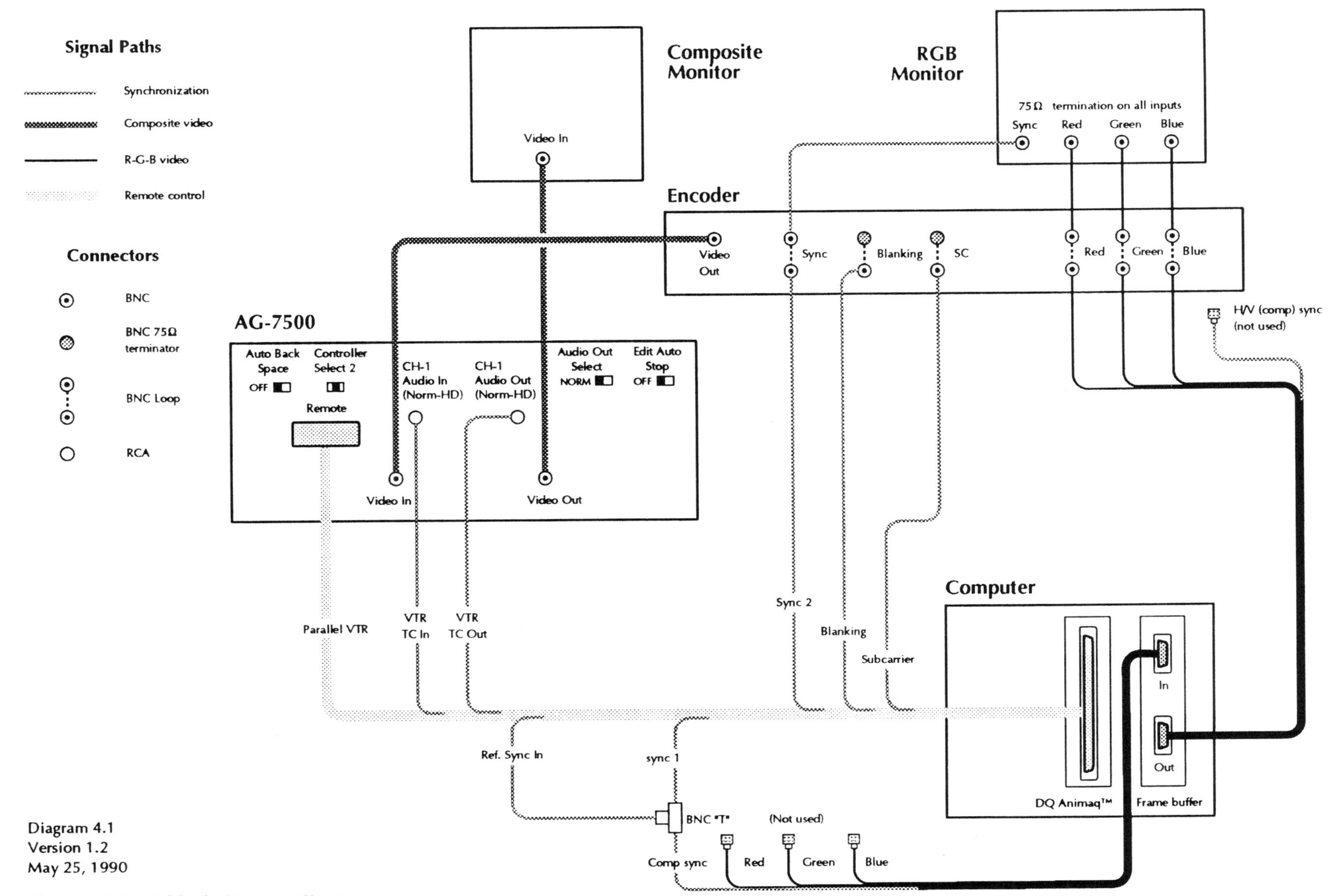

Diagram 4.1
Version 1.2
May 25, 1990

Figure 14–2. *A block diagram illustrates how a typical host/VTR/encoder/controller system can be configured. (Reproduced with permission of Diaquest.)*

Some VCR controllers have the ability to control two recorders at once. This is useful when you want to read in a video frame for image capture, render another image over it, and save the composited image onto another recorder. Also note that more advanced controllers have the ability to manage all the requisite signals for a completely professional broadcast video system; that is, they include features such as advanced sync, black burst, and other sophisticated signal interfaces.

Time Code

The controller is oriented with reference to the tape by way of the time code: Because of time calibration, the controller knows the current position on the videotape, when to start a preroll, and where exactly to place the next frame. Time code is fundamental to computer animation. It's utility is manifest throughout the video industry because:

- Time references must be precise. Information regarding length of video programs and elapsed time must be constantly available.
- It allows tape recorders to operate in complete sync. Video information can be transferred accurately.
- Video information can be accessed repeatedly, as needed.

Two ways have been developed to put critical timing information on video tape. The most common method used is to place coded sounds on one of the two available audio channels (usually channel 2). The first generation signal in Figure 14–3 shows what the signal looks like when

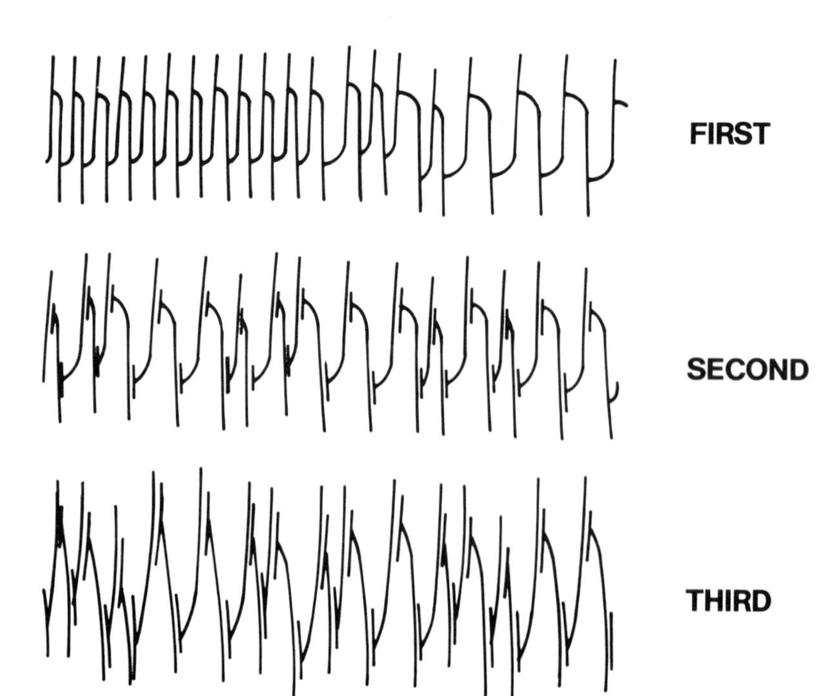

Figure 14–3. *Making multiple generations of a videotape successively increases the distortion level of the SMPTE time code.*

viewed on an oscilloscope. This code, established by the Society of Motion Picture and Television Engineers (SMPTE) and set up to identify the hours, minutes, seconds, and frames of the tape, is a standard code used in virtually all editing and production studios.

The other time code standard doesn't use either of the audio tracks on a standard videotape; instead it relies on the sync timing between video frames, called the *vertical interval*. This method of time code is called vertical interval time code (VITC). An advantage of this method (over and above the freeing of an audio channel) is that it indicates more accurately what video field (one half of a video frame) the VCR is sensing. Another advantage is that the VCR can always be oriented to tape location even while holding on a single frame. In contrast, audio-based SMPTE time code requires the tape to shuttle back and forth over the heads of the VTR in order to read the audio signals. Still, despite the limitations of audio-based time code, it is by far the more prevalent time coding method.

Formatting a Videotape with Time Code. Just as floppy disks must be formatted before they can be used, so videotape must be initiated with time code. Straight out of the box, videotape lacks SMPTE time code on either of its audio tracks. Thus, before single-frame videotaping, the tape must be *striped*, or provided with time code on the audio track along with a video signal.

If your VTR controller doesn't incorporate an internal time code reader/generator as one of its features, then an external unit must be used to put the time code onto one of the audio tracks. A blank video signal (video black) is recorded along with the time code on each tape prior to its use in production.

A few features are of special interest when you look at time code reader/writers. So called *jam-sync* capability allows you to use another time code generator to resynchronize the time code already on a videotape. Another feature, tach pulse operation, automatically and continuously maintains time code information during drop out periods, when there is temporarily no information on the tape, due to tape flaws. However, because most of the extra features usually relate specifically to applications peculiar to the video production industry, they are not discussed here.

Striping a Video Tape. With an SMPTE time code generator, it is common practice to put the time code onto audio channel 2. The level of the signal is set to 3 decibels above the 0 reference level on the VTR's audio meter. This ensures a correct signal level that can be played on other VTRs, many of which vary in audio playing level. The automatic gain control (AGC) should be off to prevent potential distortion artifacts. Finally, at least 20 seconds of black header should be allowed. This insures sufficient time for most VTRs to reach a stable speed.

Videotape Recorder Formats

Before selecting a VTR, you will need to make some determination regarding video format—another topic that can spiral into unending evaluation. You need to choose a format that properly matches your equipment.

From the number of tape formats available to the animator, you must select the one that meets the technical and/or financial requirements of your usual animation. The following is a short description of the popular VTR formats that are suitable for animation production. Some formats are not discussed here: Although they may fit into a low-cost/high-quality category, some important factor precludes their use, such as the lack of a controller or undependable single-frame accuracy.

Because of the ongoing video format war (based on cost, convenience, and technology), evaluation of appropriate formats can be difficult. Indeed, for the animator who must operate on a tight budget, and thus needs to identify suitable low-cost equipment, the options can be dismaying. You'll discover that all good VTRs, regardless of format, are expensive.

Perhaps the most expensive broadcast-quality analog format, one-inch M, is considered outdated (by some professionals), and as a result, fine used (retired) machines can be found at bargain prices. However, this discussion focuses primarily on the features of the constantly evolving lower-cost VTR equipment. The discussion starts with the least costly format and goes on to the most expensive one.

VHS. The least costly recorder format that is suitable for serious animation work is VHS, which uses standard 1/2-inch tape. Both JVC and Panasonic make high-end VHS VCRs with cast aluminium frames that make for stable, reliable operation. The problem you may run into here is finding a suitable controller. Currently, there are only a few controllers that support these recorders because of the format's inherent weakness when it comes to reliable single-frame control.

Another drawback to be considered when contemplating the use of VHS is that the resolution is only 240 lines and the color bandwidth (and hence quality) is limited. Thus, the level of quality, especially in multiple generation copies, makes this format acceptable only for low-quality or first-generation animations.

Super VHS. One level up on the scale of format quality is S-VHS, also called S-video or Super-VHS. This format offers color quality comparable to VHS, but the resolution is significantly enhanced: S-VHS generates better than twice the black and white resolution of VHS.

S-VHS requires special tape with higher coercivity (magnetic density). The S-VHS VTR is downward compatible with VHS—in other words, it can play back or record in VHS mode—however, S-VHS tape cannot be played on a VHS deck. S-VHS equipment utilizes a different video signal and requires special connectors that are not VHS compatible. The format separates the chroma from the luminance and, as a result, offers a much better signal quality than VHS. For optimal viewing quality, you should use an S-VHS compatible monitor, which can display the video without mixing the chroma and luminance. The quality is somewhere between that of NTSC and RGB.

Make note, however, that in all probability, the viewer will see your animation in composite mode, which produces the artifacts S-VHS so noticeably removes. The S-VHS format was (and is) marketed as a low-cost alternative to the still popular mainstay, 3/4-inch format. But even though S-VHS quality is comparable to that of 3/4-inch, the ubiquity of 3/4-inch equipment and its industrywide acceptance and support represent considerable momentum, thus making it a daunting rival for the newer 1/2-inch S-VHS format. Industrial-quality editing S-VHS VTRs, also made by JVC and Panasonic, are very reliable continuous-operation machines. Also, a range of suitable controllers, TBCs, and advanced editing machines that support S-VHS machines are currently on the market.

The S-VHS format is not without its drawbacks, some of which may preclude its satisfactory use for a particular computer animation. One of these is the color encoding method. Some VTRs produce a chroma delay as a by-product of improving the quality of the color. The color processing appears as though the colors have been shifted upward one scan line. Although this may not be perceivable in a first-generation playback, it is more noticeable on third-generation copy (see Table 14–1). The good news is that a competent time base corrector or TBC (one specially matched to the deck) will correct this problem.

Based on color quality, it's hard to say categorically whether the S-VHS format or the popular, industrial-standard, 3/4-inch format is superior. Highly saturated S-VHS colors do not retain color accuracy as well as do 3/4-inch color images. Computer generated images typically possess highly saturated colors, and this can be troublesome with this S-VHS format. On the other hand, the higher black-and-white resolution of an S-VHS image makes color contrast appear clearer in low color-saturated images.

The 3/4-Inch Format. The 3/4-inch or U format has been around, and improving, for more than a decade. Its predictable performance and mid-range cost have made it popular, and its use is widespread. Despite the better characteristics available in other formats, when it comes to finishing, you can be sure that a 3/4-inch editing facility is close by. This

Table 14–1. Performance Loss From Multiple Generations of an S-VHS Deck

	First Generation	*Second Generation*	*Third Generation*
Horizontal resolution (lines)	400	370	350
Signal-to-noise ratio (dB)			
Luminance (color)	57.2	51.7	49
Chrominance (AM)	51.8	47.5	44.5
Chrominance (PM)	44.3	40.1	35.2

is definitely a consideration: Many animators need to be assured of quick and complete services after an animation segment has been created.

The 3/4-inch format uses the standard method for displaying color and luminance information. This means that you can expect to see pervasive artifacts of NTSC, such as chroma crawl and colored Moire patterns caused by the proliferation of fine black-and-white lines. Despite these drawbacks, if you use good tape and a reliable VTR, your finished quality should be acceptable for editing.

One recent improvement in the 3/4-inch lineup is Sony's Superior Performance (SP) series. Both variations of the format are compatible with one another, with the exception that SP recording requires SP-editing VTRs. SP relies on improved circuitry and tape; Sony claims that these improvements result in about a 30% improvement in overall quality. Because the resolution is enhanced and color interference has been reduced, the improvement over the 3/4-inch U format is indeed noticeable.

Betacam. The next level of equipment represents a significant improvement—an improvement that is echoed by a comparable boost in equipment costs. Sony's Betacam and Betacam SP VTRs are expensive, but they offer excellent quality. The reason for the marked improvement in video quality is that Betacam uses the component video format and an improved method for recording the signal onto the videotape. This, along with superior tape, provides for seamless dups and edits at multiple generations. The Betacam recorders, in contrast to the less expensive VHS series and 3/4-inch VTRs, use a serial port (technically, an RS-422) for the remote control of the VTR. Betacam is an increasingly popular format and, in terms of overall video quality, is considered by many professionals to be superior to top-of-the-line broadcast 1-inch machines costing in the area of $100,000. In other words, even for broadcast applications, Betacam is more than adequate.

Digital Formats. Although digital formats are growing in popularity, they are still out of the price range of most low-cost computer animation facilities. The digital formats, D-1 and D-2, have a clear advantage over other formats because each generation (copy of the original) poses no image degradation. For the animator, this means that multiple overlays and composited images can be repeatedly integrated into one image or animation without any price in terms of image degradation.

Videotape Recorders

VTRs, like VCRs, range widely in capability and cost $3000 to $140,000, so selecting one for animation use warrants careful consideration. Keep in mind that single-frame animation is particularly hard on VTRs. For every second of animation (30 frames), the VTR must be on constant standby, waiting for the next rendered frame. Then, on signal, the VTR must go into action: It turns on the drive motors, prerolls, plays, records a frame, stops, and goes into operational limbo again, waiting for the next rendered frame. Therefore, if preroll takes 15 seconds, 1 second of rendered animation represents about 7 or 8 minutes of stop-and-start VTR operation. This is a mechanical strain on the hardware—wear and tear will take its toll even on the stoutest equipment, particularly on the video heads, the components that directly contact the videotape.

VTR Selection. Computer animators always seem to be in the market for a new VTR, because it is such a critical component. You will find that, over and above determination of format, selection is not made easier by the continual introduction of "improvements" and by the attempts of manufacturers to launch new standards as they "update" features (see Figure 14–4). Given such a fluid marketplace, no guide (including this book) is adequate to the task of asssessing the latest VTR capabilities. For up-to-date information, consult the video magazines for

a

b

Figure 14–4. *The features in low-cost videotape recorders are ever increasing. (a) The Panasonic unit, for example, includes an internal time base corrector. (b) The NEC PC-VCR is unique because it was designed with a serial port that connects directly to a computer. ([a] reproduced with permission of Panasonic, [b] reproduced with permission of NEC Technologies, Inc. Professional Systems Division.)*

trends, and ask professionals at video facilities about the benefits of units currently on the market. You are not likely to find much consensus regarding a unit's technical virtues, but the process will sharpen your awareness of what's available and make you more adept at cutting through the jargon and sales hype.

Of course, the VTR you select must not only support your particular format and animation needs—based on features currently available—it must also be affordable. Indeed, you may discover that, depending on the quality of the animation you wish to produce, this is a tall order, especially if you are also shopping for durability.

Maintain Your Tools. To insure dependable quality, VTRs must be maintained and serviced on a regular basis. VTRs are mechanical devices and so are vulnerable to time and dirt. Clean the heads often with high-quality cleaners, keep the units well ventilated for cooling, and avoid dust. You should clean the heads properly and carefully and make note of how many hours the VTR is on and for what purpose. After a manufacturer-rated time, the VTR should be given to a professional for checking and calibration. You may not notice degraded performance in a first-generation animation, but you can be sure that it will be noticed as multiple generations accumulate. Finally, set up your VTR in a cool, well ventilated, dust-free environment.

Magnetic Tape. Regardless of which VTR you use, the use of high-quality tape is essential to quality animation. Each frame of an animated sequence requires repeated stopping and starting of the tape's transport mechanism, so the tape must endure repeated stretching, hitting, and flexing and still maintain its specified physical characteristics. That's why, as a general rule, tape used for animation is used only once. If you use it again, be prepared to find dropouts, caused by the flaking of tiny magnetic particles embedded in the tape. Dropouts appear as small but momentarily visible on-screen blemishes.

When you are shopping for a VTR to be used for animation applications in your chosen format, some key questions can help you elicit information you need about relevant capabilities:

- Can the VTR handle continual operation? Remember, it will probably be in operation 24 hours a day.
- Is the unit consistently frame-accurate?

- Are the heads designed to withstand the countless prerolls and standbys that characterize single-frame animation?
- Is the format compatible with available editing facilities and your usual client's preferences?
- Is output quality (including of multiple generations) acceptable?
- What professional features does it offer? Do you need them?

RAMcorders

Digital recording on a device such as as RAMcorder (for real time playback) is an excellent choice for the animator who can afford one. RAMcorders, such as an Abecas A-60, digitally store images on fast hard disks for real-time playback. These recording systems offer many advantages over mechanically oriented VTRs. One of the most obvious improvements is that there's no preroll to deal with. This saves 15 seconds per frame, as well as the waiting that occurs during the VTR's starting and stopping.

A secondary benefit is that images can be stored digitally; thus they retain their pristine original quality. Another advantage offered by the RAMcorder is that it can be played forward or backward at any speed. Finally, with hard disk storage, there are no tape artifacts to worry about. Some of the high-end units, such as those made by Abecas, also enable a wide range of special effects.

Video Disk Recorders

Optical media have been economically out of range for most animators; yet the price is dropping toward competitive levels. Rewriteable optical disk recorders, like those made by JVC and Panasonic, cost in the area of $15,000. A single disk provides 54,000 video frames in either high-quality RGB, S-VHS, or standard composite video. Further advantages of optical recorders are that the units do not require preroll and their lack of complex mechanics translates into a long lifetime.

Video Editing

Video editing is a very different process from film editing. Indeed, the subject of video editing deserves a complete book unto itself and thus will not be covered in detail here. A number of significant advances have recently been made in the video editing process, particularly relative to desktop computers. Although not all of these will be discussed here, a few important video editing aspects that are common to most editing sessions will be covered.

Regardless of what editing software you use, a list of what section was positioned where and how that transition was achieved is important and necessary documentation. You will find that it's especially important when it comes times to modify your work. This listing of times and events is called an *edit decision list* (EDL). The EDL can be saved on disk (in a standard format); if it was produced using any of the popular desktop video editing programs, the listing will be available for use by other programs. For instance, it could be used by a commercial video facility for automatic assembly (using the original videotape), resulting in a very high-quality end product.

Professional editing is done in two steps: off-line editing and on-line editing. *Off-line* editing includes the process of creating the EDL. Here, duplicates of the original video material are used to create the edit points. After the EDL has been created, *on-line* editing procedures serve to finish the edited master videotape.

When a video is produced, two basic editing procedures predominate: assembly editing and insert editing. *Assembly editing*, as the term implies, involves the simple addition of a video sequence onto a preceding one (see Figure 14–5). Thus, assembly editing is a sequential operation and is the usual method of grouping the parts of a production. This is how most animation sequences are created.

Insert editing is more complicated: It involves the insertion of new video information into previously recorded material (see Figure 14–5). Two VTRs are used; one serves as the source, and the other serves as the target. When a sequence is to be inserted, the source material replaces selected sections of the target's video material. As you might suppose, timing is critical to the operation, and so it is necessary to use a good VTR so that frame-accurate editing is assured. Prior to the operation, the target tape must be completely prerecorded with a video signal. It's otherwise impossible to insert video signals where there is no pre-existing video information already on tape. Some editing systems facilitate the insert procedure by providing a preview mode, which allows you to view the changes you anticipate before you actually record them (insert them) on the target video.

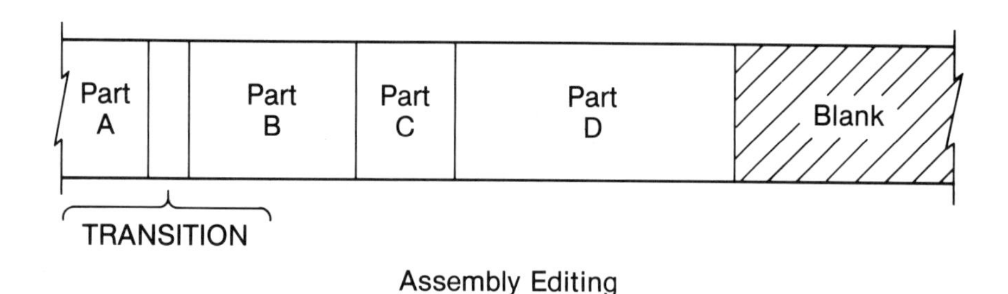

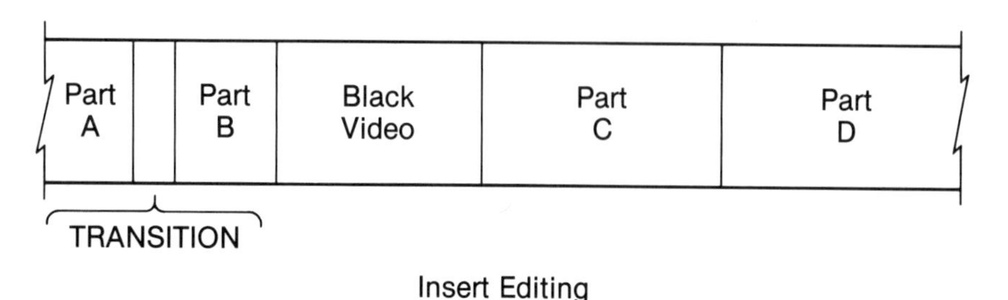

Figure 14–5. *Assembly and insert editing differ based on the editing process. Assembly editing is segmented, adding new video material to the tape. Insert editing, as the name implies, allows you to insert video material in a pre-existing video sequence.*

A/B editing is a video term describing the routine of switching (usually a dissolve) from one video sequence to another, a process that necessitates a video editing system (see Figure 14–6). This seemingly simple technique requires a somewhat complicated equipment setup—three VTRs are needed. Two of the machines are dedicated to the source images, and the other is used as the target. After the edit points have been selected and the dissolve time established, the two source VTRs are set to preroll and play. After they attain the proper speed, the preview screen displays the smooth transition from the source of one VTR to the source of the other. Finally, the third VTR records the result.

Editors that possess these control functions are called *video switchers* because they simplify the control and switching of the video signals to and from different devices. Switchers can be simple, relatively inexpensive devices, or they can be technologically complex units (costing in the six figures) that incorporate many extra features to facilitate video editing. Some video switchers may include such features as the control of fades (to black or any color), control of the mixing of different video sources, special effects, and interfacing to a wide variety of devices. Units designed for professional applications (usually found in a video production facility) are extremely complex in operation and so are beyond the realm of the desktop animator. In some cases you'll find that it's best to turn over finish editing to an operator familiar with a particular unit's capabilities.

Transitions. An animation (as well as almost all video production) typically is composed of several sequences that need to be assembled. In such cases, after the various segments have been created, *transitions* must be provided to smooth the flow of the visual process from one scene to another. Transitions can be accomplished in a number of ways using a variety of tools. Because the transition method selected invariably affects the mood of your work, you need to plan your transitions. You will quickly discover that transitional devices are important elements in a kind of visual vocabulary and that they color the meaning of your visual text. You'll also find that the transitions you use have explicit denotations and surprisingly subtle connotations.

The simplest transition and the most straightforward is the *cut*: One sequence is simply commenced immediately after the conclusion of another. From an editing standpoint, this abrupt shift is easy to manage, but it may be hard on the viewer. A cut may be perfectly appropriate, or

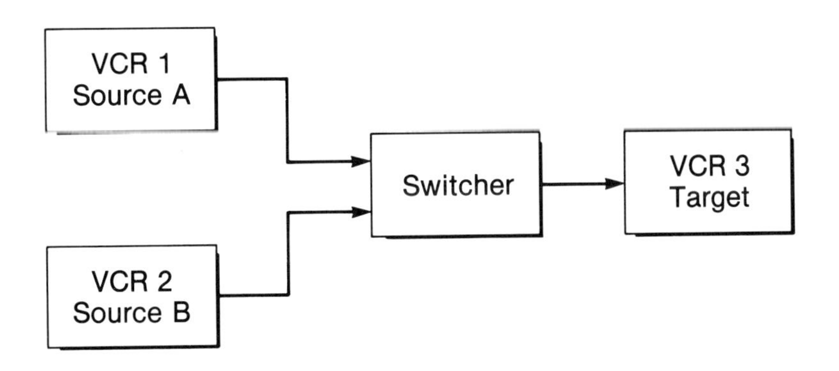

Figure 14–6. *A/B roll relies on three VTRs, two source decks and one target. Using this arrangement, a smooth dissolve from A to B can be created.*

it may appear harsh and precipitous in the context of the animation, especially when there are a lot of them in sequence.

Another common transitional device is the *fade*, usually either to black or to white. This transition is typically used to conclude a sequence. A quality of finality is associated with it.

A *dissolve* is a smooth, gradual transition moving from one sequence to another. It is a much gentler transition than a cut. The length of time required to accomplish the dissolve is arbitrary (determined, that is, by the editor) and can be managed on the computer as a part of the animation. A dissolve can also be superimposed after the fact via an editing system.

Certain animated camera functions, such as the pan and zoom, can also be used as transitional devices. For example, a sequence can be set up so that the camera will pan into a new and different sequence while retaining a uniform pace and direction of movement.

These simple examples only begin to suggest the unusual animation capabilities available to the desktop animator. You will discover that the computer can provide a wide range of interesting and novel transition options (see Figure 14–7).

Effects. An often overlooked animation technique involves the integration of special effects. Even though advanced video post-production facilities can initiate special effects at the press of a button, numerous effects are not available to them. In other words, despite the elaborate equipment and facilities, they are limited in some ways that the desktop animator is not.

A number of computer generated effects can be integrated into the rendering process instead of during post-processing. For example, an image can be rendered as a background and then, via post-processing, it can be made to appear out of focus by blurring or averaging the image after it has been rendered.

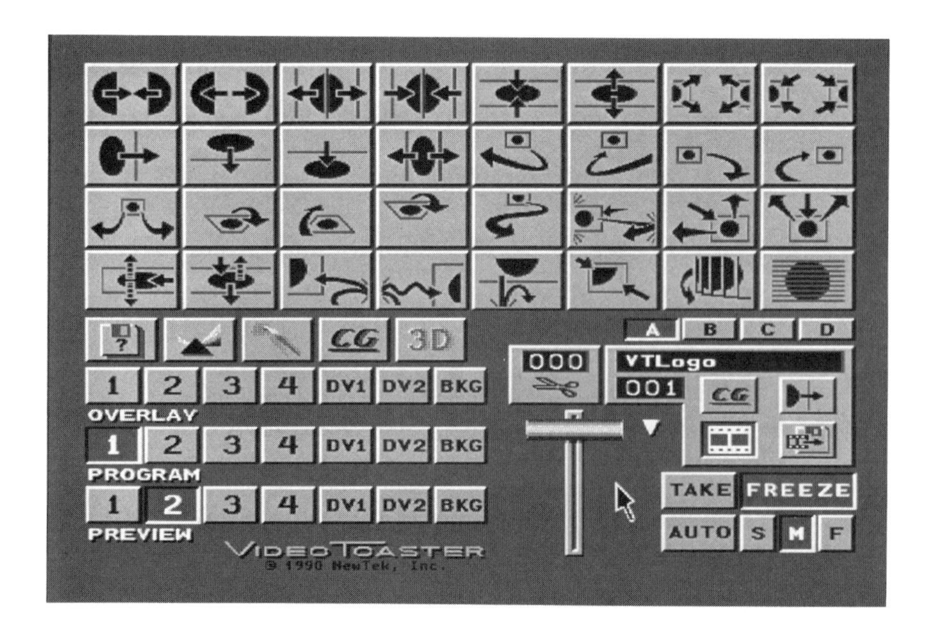

Figure 14–7. *Effects for transitions can be chosen by software that mimics the consoles of more expensive hardware versions. (Video Toaster software/hardware reproduced with permission of NewTek Inc.)*

The same effect can be achieved by way of video. For example, a live video sequence of poor quality can, through image processing, be enhanced to the point of acceptability. In fact, some post-processing programs have the ability to create what are called "painterly" effects. Using these programs, you can create images that appear as if they were drawn with pastels, charcoal, or pencils. The resultant images can be used as is, or they can serve as backgrounds for a composited animation.

15 Recording to Film

Compared to video, animating with film as a target medium provides for some dramatic contrasts. There are stark differences between these media in terms of both rendering and viewing. The technologies are dissimilar, and so is the viewing experience. The impact of image size cannot be overlooked: Film is typically viewed in a significantly larger format than that of a video display. Therefore film demands greater quality in detail. Again because of the scale, you'll find that defects on film stand out larger than life. Flaws that would be insignificant on a video display can be obvious, even impressive on film. Fortunately, however, you'll also notice that the impact of at least some defects is diminished by the potent scale of the viewing experience. Even though the differences between video and film technology may appear obvious, the following discussion takes a more scrupulous look at these differences as they effect the production of a well-made computer generated film.

One of the major differences between film and video is the frame display rate, how many frames per second (fps) are displayed. Video operates at 30 fps, whereas film is displayed at 24 fps. The difference in rate engenders built-in compatibility obstacles between the two media; you will find these obstacles particularly problematic when it comes to editing.

Another major difference is purely visual: An astute observer can detect slight frame differences between the two media even when they are viewed in an alternate format (for example, viewing a film on TV). The same astute observer might also notice, while watching a video, telltale flaws that signal its film origins, such as an occasional scratch or a persistent mote of dust—a clear sign of film's more physical nature (and of poor quality control, too).

Film buffs will also point to video's inherent limits in terms of dynamic range, that is, in its ability to convey full darkness and light. Film's dynamic range is significantly wider, making for a much more impressive overall image. Standard video offers a usual contrast ratio of between 15:1 and 30:1, whereas film extends the contrast ratio to over 100:1. The extended visual power of film allows the desktop computer animator to craft highly dramatic scenes which convey considerably more richness and depth than can be found on video.

The color chart (see Figure 13–3, Chapter 13) demonstrates the limits of composite video with respect to the visible color spectrum. You will find, for instance, that pure red or aqua on video will lack saturation value and will instead be replaced by a paler representation than seen on the original RGB display screen. Thus, a saturated red will become instead a bright pink. In contrast, film provides an incredible range of rich color saturation, a characteristic well matched to the color capability and range of computer graphics imagery.

The best resolution of 16-millimeter film (as measured by its acutance or granularity) is greater than that obtained by even the highest quality NTSC video. The resolution of video is, at best, about a few hundred lines, in contrast to 16-millimeter film's 1800 lines of resolution. In terms of resolution, 35-millimeter film is significantly better: It is strikingly clear because it resolves over 2500 lines. These are static measurements, however. When animated, a film's resolution appears even better. The tiny grains endemic to each frame of film are rapidly replaced by a new set of tiny grains from the next frame, and this effect reduces the appearance of granularity.

Another advantage film offers the animator is that it is a relatively stable medium, and so it is better suited for archiving. Over time, videotape is susceptible to flaking of the magnetic particles that electronically store your animation. There is also the problem of constantly changing video formats. Many videos made decades ago cannot be played because the VTRs are no longer in existence. Regardless, the video format is not considered a long-term medium given the present state of the technology.

In contrast, when an animation is stored on film, you can expect it to retain its essential characteristics for decades, and, when necessary, it can be transferred easily to whatever video format is currently popular. This is one reason cartoon producers usually make their original animations on film. They must plan on the likelihood of (profitable) reruns, which inevitably require high quality dupes. They also know from previous experience that foreign distribution will necessitate the transfer of the animation to different, sometimes unpredictable, video formats.

A disadvantage of film is that editing is somewhat more difficult: Most professional animators consider video easier to work with. Video also allows easy access to a fuller range of special effects and other features. However, if the original material is created on film and then edited in video format, the animator can enjoy the benefits of both media.

Production time is an important issue that highlights a major difference between film and video. The exposure time for an analog film recorder (discussed later in this chapter) is about 5 to 15 seconds per frame—a time penalty similar to that incurred with video preroll. However, the required exposure time for digital film recorders (also discussed later in this chapter) may be as much as 15 minutes per frame!

As you might imagine, at 4 frames per hour (32 per 8-hour day), production time can overrun a budget in a hurry.

When working with film, another concern is quality control over film stock. Due to temperature changes or production variations in the manufacture of different film lots, a particular roll of film may be so noticeably unmatched in color balance that a reshoot is necessary. Such unwanted surprises can be hazardous to quickly approaching deadlines.

If you are working in the film format for eventual viewing on video, you should anticipate the eventual translating process limitations. For example, if your animation's colors (on film) are heavily saturated, the colors of the end result may be disappointing.

Here's one last caution for those who plan to work with film: Be advised that even the best transfers of film to video produce artifacts, often very noticeable ones. You need to work closely with the transfer operator so that you can be assured that the finished quality of your animation meets your standards.

A Low-Cost Solution: Shooting from the Screen

Shooting animation sequences with a camera directly from the CRT screen is the least expensive transfer method and can produce acceptable amateur results. You can expect the quality of a colorful animation produced on film to be degraded because of the nature of the color CRT screen. The metal mask inside the display monitor that defines the positions of the color dots is as visible as the computer graphics. In fact, it sometimes interferes with the computer generated image by causing Moire patterns (this is especially true with Trinitron or lenticular CRT displays).

Another common problem caused by shooting from the screen is geometric distortion. Geometry problems can make a square appear bowed, slanted, or concave. You can correct many of these unwanted effects by using a telephoto lens. The long lens, in effect, distances the camera from the screen, reducing the appearance of perspective. This loss in perspective also makes the bulb-shaped display screen appear flat. Not everything is correctable, however: Many displays produce geometric distortions of the image that may be impossible to correct.

It is not difficult to control the camera via the computer if you are prepared to build simple electronic and mechanical controls yourself. A hobbyist–technician can design and construct a low-cost film–camera relay that is controlled from the computer speaker or serial port, which will activate the camera for each frame. However, as for any custom arrangement, many test trials will be required before the system can be relied upon.

Commercial Film Recorders

Most film recorders that are used for still or motion picture film are analog in design; that is, they use the RGB video signals from the graphics display system in order to expose the image onto film. A digital film recorder, in contrast, relies on digital information, often sent via a parallel port, to reconstruct an image on its own internal display system (see Figure 15–1).

Incidentally, this discussion of film recorders refers as well to units that produce single computer generated images. Film recorders for animated films are identical to units designed for single images. The only difference is the type of camera that's used.

The main advantage to the use of a commercial analog or digital film recorder (over shooting the image from the CRT screen) is that reliable and repeated exposures can be automated. Film recorders are manufactured and set up for use with specific film types, and when they are properly employed, they yield dependable resolution and excellent geometric linearity.

Digital film recording is the most expensive method because of the length of time required to expose a single frame and also because the equipment is more expensive (than its analog counterparts). There is, however, a significant payoff in terms of quality. Very-high-resolution images are generated; this is why digital is the method of choice in the motion picture industry. Analog film recorders are limited to about 1400 lines of resolution, whereas digital film recorders start at 2000 to 4000 lines.

Here's how they work: The digital film recorder reads the image data output from the computer through either the serial port, the parallel port, or a custom interface board. In other words, the computer outputs digital information instead of images. These high-resolution images, even when compressed, can consume as much as several megabytes per frame. The recorder displays and records the reconstructed image on the film recorder's picture tube according to its instructions. The digital format of the images is dependent on the particular film recorder, so you should plan ahead for file compatibility if you elect to use a digital film recorder.

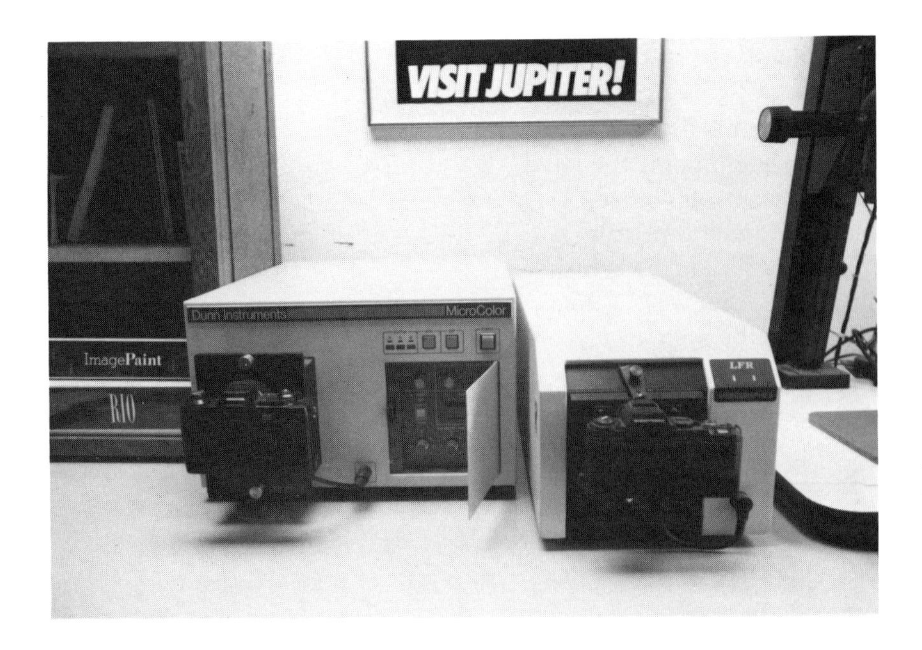

Figure 15–1. *Although similar in appearance, digital and analog film recorders produce different results. On the right is the LFR digital film recorder from Lasergraphics, and on the left is the Dunn Instruments (now Matrix) analog film recorder.*

In contrast to digital recording technology, analog film recording work is based on a relatively straightforward scheme. An enclosed box houses a black-and-white CRT at one end and a camera at the other. Between the camera and the CRT screen is a rotating color filter for the red, green, and blue signals. Rather than projecting a color image, the system is set up to display only one color at a time. This approach addresses one of the weaknesses of a color monitor's display: The color mask inside the picture tube, which is always visible, decreases resolution. Additionally, monochrome monitors possess higher resolution while costing less. The colored images are exposed sequentially as the filter is rotated in front of the lens. In this way, the film is exposed to the correct color image, one color at a time.

Almost any 16-millimeter or 35-millimeter motion picture camera will work, providing it can expose a single frame at a time and can be properly mounted to the film recorder (see Figure 15–2). A small motor controls the execution of the exposure and the film winding. Well maintained equipment is especially important here. A common defect of film recording, seen even in major theatrical releases, is called *film or tape weave*: The image appears to move slightly yet gradually about the screen. Tape weave is caused by slack in the camera's film holder, which is supposed to exactly line up the film in the right position for exposure. When you see this artifact, you know that you are viewing a computer generated animation on film (even though you may be seeing it on video). For exacting applications, therefore, such as matching another animation or for precise video-to-film registration, a pin-registered camera is used. *Pin-registration* is a mechanical solution to the mechanical problem of tape weave. After the film is properly seated for exposure, small pins are snugly positioned into the film's sprocket holes, providing an accurate and repeatable film placement. Some cameras without this feature can be retrofitted with pin registration.

Figure 15–2. *A 16-millimeter camera is used with this analog film recorder for automatic exposure on motion picture film.*

Because of inherent color differences in film types, which are especially noticeable with longer exposures, the time for each red–green–blue exposure is different. Luckily, commercial film recorders are preset to accommodate the different exposure times needed for the more popular film types. Also, adjustments in these film recorders provide a means for altering the colors for custom applications.

Communications Compatibility. Analog film recorders are usually controlled by way of the computer's RS-232 serial port. This is how you communicate with the recorder when you want to modify the settings or command an exposure. If you are having problems with your recorder, check to see if the computer's communications parameters for the serial have been set correctly. For MS-DOS computers, for example, use the MODE command, which sets the baud rate, parity, stop bits, and so on. It's possible, for example, to incorrectly set the port to 7 bits. If the film recorder is expecting 8 bits, you could have sequencing problems. The film recorder might execute the exposures, occasionally missing a frame.

Regardless of film recorder type, film stock selection is limited in sensitivity and type, so film parameters should be carefully evaluated. Fortunately, very fine grain film (with the lowest ASA or film speed) can be used because the light level and exposure are controlled by the film recorder. Negative film stock is better than positive transparency film because it has superior grain characteristics, the latitude of exposure is greater (providing a wider range of "under and over exposure errors"), and the contrast is lower (providing a wider choice in final image quality). Film editing is usually done on a *work print*, which is a positive of the original negative. For applications where video is the target medium, the original negative is transferred to video for editing.

Shooting on positive film can also produce acceptable results, although you should do this only if you have a very limited budget or want to create a preview of your animation. Luckily, positive film stock is thicker and more durable than its negative counterpart. This is an important consideration if the editing is to be done on the original positive (which is not recommended for quality work). However, you need to keep in mind that if the contrast is set too high or if any other attributes need to be changed, the exposed film can't be modified; it must be reshot.

Film-to-Video Transfer

Transferring the negative (or positive) to video is a complex process usually relegated to larger systems such as those made by Rank-Cintel or Bosch. Each system has its own special attributes, quirks, and features. Before transfer, the operator needs to know at what rate the film was shot, 24 or 30 fps. At 30 fps, the video-to-film frame ratio is directly 1:1, and in this case you can avoid any compensation for time differences. Another benefit of the higher than normal film rate is that it decreases apparent grain, making images look sharper. A disadvantage, of course, is that more frames must be rendered.

Other video issues to consider when transferring to video are granularity, exposure time, quality control over film stock, and the generation of artifacts as a result of the transfer onto video. Surprisingly, granularity, inherent in all film, can be beneficial to the animator because it lessens the effects of flickering of single horizontal lines and tends to reduce aliasing artifacts during transfer.

Also keep in mind that the NTSC aspect ratio of the screen image (height to width) is 3:4. Because of this, your film recorder should be adjusted to expose the film on the basis of this ratio. Finally, remember that home TVs typically don't display a full video image; thus the rendered image should be within the safe titling area, which is 80% of the total video image.

Plan to work closely with the operator of the transfer system to help ensure qualilty. Many subtle decisions are made in the course of the transfer. For example, because film contrast is invariably compromised when images are ported to video, you may have to decide whether to expose for the highlights or for the shadows. Additionally, the operator, called a *colorist*, can tweak any of the colors you wish, including matching the colors of another film.

If the film is shot at 24 fps, a different shutter blade is used on the projection of the film-to-tape transfer system. The process of converting film to a reduced video rate is referred to as a *3:2 pulldown*. The pulldown process works like this: Periodically, a frame of film is exposed for a video duration of three fields (1.5 frames) instead of two fields. For the first film frame, the projector exposes the video to two fields; on the second film frame three fields are used—hence the term 3:2. Newer transfer systems incorporate variable speed transfer rates because they can scan the film one line at a time. This results in smooth transfers for various film rates. It also means that the video transfer system operator can slow down or speed up the original film rate as required.

As you might expect, the 3:2 pulldown causes editing problems on the video side. Edit points may include frames of different fields so that, for example, wide pans might appear somewhat jerky. However, careful editing on a good system can remedy most of these problems.

Putting It All Together

16

Using a Service Bureau

As you may have surmised, producing video can be tricky and is usually expensive. Supporting the various VTR formats that your clients may require can be prohibitively expensive (or at the least, impractical). For example, the cost of a better tape recording deck starts at $100,000. Even if you amortize the cost over several years, you need a generous cash flow to justify such an expense.

That's why the use of a service bureau is a sensible alternative for many animators when it comes to production of the end product. Most video post-production facilities (a place where final video editing is done) provide their services at reasonable rates. They know that, after an animated product has been rendered, they are the most likely option for finish editing, duping, and the inclusion of special effects.

There are two ways you can render an animation onto tape (off-site). The animation can be already rendered (by you) as a sequence of images, or it can be submitted to a service bureau or other facility as two files, one of the 3-D models and another of the motion. In the first method, all digital images are stored on some form of removeable mass storage for the service bureau to read the sequence of images and put them individually onto videotape. Keep in mind that rendering onto a file for transfer to video is a notably memory-intensive procedure. If each image requires, say, half a megabyte (as it does for uncompressed Targa-16 images), then 10 seconds—300 frames—necessitates about 150 megabytes. Although this is clearly a considerable chunk of memory, 150 megabytes can be stored readily on a streaming tape or DAT format.

Using the second method, you transport your model and script data to a service bureau for rendering. The bureau then puts your animated sequence onto videotape at their facility. The advantage of this approach is that you can focus on the creative aspects of an animation and leave the mechanics and time constraints of rendering and video production to others. You liberate storage and can work on new animation projects because the computers have been freed up: You aren't waiting for rendering to finish.

In this scenario, you need to generate two types of files. First, the 3-D model (along with any 2-D backgrounds or texture maps) is put together as a model file. Second, an animation script is created; this script describes the keyframes and the motion of the objects. This information can be delivered via diskette or transmitted by modem directly to the post-production facility. Most facilities have on-site all support equipment needed to ensure the level of quality intended (up to and beyond FCC's standards).

Whatever method you use, check that your hardware and media are completely compatible with the equipment available at the service bureau you select. At the same time, you will need to schedule: Does the facility have enough resources (people and time) to do your rendering and editing within the time frame you have projected for completion of your animation? Planning ahead with a service bureau will head off problems that, otherwise, are almost predictable.

Completion Details

When an animation is completed, a number of details still remain before you can get on to other tasks. Be sure to make well-labeled backups of the models and animation scripts and keep them in a safe place. These can be used as a library for future animations. In the event that a still image is required (for, say, a promotional poster or insert) your models should be easily accessible for rerendering at higher resolution. Or you may want to make still images of special frames, which you can use in a portfolio. Keep the labeled master video or film in a very safe place, and store duplicates in a different place. Doing it later is much less interesting than doing it now, so make the duplicates before moving on to your next project.

It's also a good idea to keep notes on how long each part of the animation took. When you need to put a price tag on the development of a model or to figure overall costs when bidding for an animation job, you can check back and determine realistically how long each procedure required.

Final Thoughts

A finished animation must be put together in such a way that it can be viewed as congruent and complete. The parts should flow together to form a stylistically consistent whole. As you plan the computer generated components, you also need to consider the editing, (wipes, dissolves, and so on). Editing and transitional elements need to be carefully integrated into the animation process. In other words, it may be more appropriate (and less costly) to create some of your editing effects before the final edit so that the concept and timing of certain transitions is complete beforehand. For example, you need to provide enough lead-in of static images for long dissolves if further editing is to be done later, after the animation sequences have been completed.

The development of your animation skills is a function of both your technical experience and your communication experience. Your development is a continuous process. Computer animation is, after all, a communications vehicle. It is worth your while to find out not only what you intended to say but what was actually communicated to your audience. When viewing your animations with others, listen to the comments and opinions your work elicits. The commentary will certainly surprise you. The gratuitous opinions, insights, omitted observations, and emotional reactions are invariably revealing and can provide you with insight into more effective use of your tools and skills.

Appendix: Computer Animation Setups

The following descriptions are examples of typical computer animation systems based on the three popular personal computers. Each system represents particular needs as indicated by the choice of the components. These examples are only guides. Your needs will determine what hardware and software choices are best for you in addition to meeting your budget.

System 1

This system represents a complete system that has the capability of creating top-of-the-line animation for broadcast production. Note, however, that there is no video test or support gear, such as waveform monitors, on the list. These items may be added as needed.

Hardware

Host. IBM PC/AT (Micronics 33 mHz '486), Sigma VGA graphics board, 13-inch Sony 1304 monitor, 190-MB hard disk, 8 MB of RAM, streaming tape drive for backup.

Graphics System. Truevision Targa-32 Plus board, Truevision Vid-I/O converter, 19-inch Mitsubushi (HL6915) color monitor.

Peripherals. Wacom 9-inch graphics tablet with stylus and puck, 700-watt battery backup, JVC RGB video camera for input.

Output. Lyon-Lamb miniVAS video controller, Sony BVU-5850 3/4-inch VTR, composite monitor for previewing.

Software

3-D Studio from Autodesk, Lumena (Time Arts) 2-D animation and paint system. Inscriber (Image North Technologies) for video titling and effects.

System 2

The second system is Macintosh-based and is also capable of creating broadcast capable video animation. It uses S-VHS video equipment, which costs less than 3/4-inch video gear.

Hardware

Host. Macintosh Quadra with Graphics Accelerator, 19-inch SuperMac monitor.

Graphics output. Truevision's Nuvista board.

Peripherals. 500-watt battery backup, optional JVC RGB video camera.

Output. Diaquest video controller, Panasonic AG-7750 S-VHS VTR, S-VHS monitor.

Software

MacroMedia's Director for 2-D animation and their 3-D animation package. An alternate 3-D package is Sculpt 4-D from Byte-by-Byte.

System 3

This last system costs the least and includes the Video Toaster from Newtek which offers video switching and special effects.

Hardware

Host. Amiga 2500 with 4 MB RAM.

Output. Sony Hi-8 VTR, standard composite monitor and RGB monitor.

Software

Lightwave from Newtek.

Glossary

ADO: Ampex Digital Optics. Trade name for Ampex's video special effects generator that replicates opticallike effects.

Alias artifact: A visual defect as making a diagonal line appear jagged or staircased, caused by the limitations of the computer's display resolution.

Alpha channel: An additional capability on graphics boards that extends the red, green, and blue outputs and provides functions such as video, transparency, and matting.

Ambient light: Lighting that illuminates a scene from all directions, such as the lighting on a cloudy day.

Analog signal: A signal whose amplitude can vary from one value gradually to another.

Anti-aliasing: A mathematical process that makes stair-cased diagonal lines appear smoother by averaging (filtering) the surrounding colors.

Artifacts: Unwanted results from a process, such as imperfect shadowing.

Aspect ratio: A number set representing a display's height to width. For video, the aspect ratio is 3:4.

BBS: Electronic bulletin board system. Like an electronic mail box for storing digital files via a phone, modem, and computer.

Bicubic patch: A four-edged 3-D surface that is a mathematical representation instead of a flat polygon.

Bounding box: A boxlike representation of an object, often created to facilitate the object's movement or to provide simplicity when viewing other objects.

Bump map: An image used for mapping onto an object to give the appearance of bumpiness on an otherwise smooth surface.

Bus: The socket or electrical connection between the computer and the plug-in boards that work with it.

Cartesian coordinates: A mathematical system of graphing the x, *y*, and z values.

Chrominance: In video, the color part of the signal.

Chroma crawl: An unwanted artifact from NTSC video that appears as moving lines when two colors (such as blue and red) are next to each other.

Chroma key: A method for "keying" or locking onto a color so that it can be used to display video information.

Color look-up table: Memory that holds particular assignments of color values.

Color purity: The closeness in quality of the color display.

Component video: A video signal whose color and black-and-white information is recorded separately, thus making the video quality superior to composite video and improving the quality of multiple generations of videotape.

Compositing: Layering of various graphic images together.

Composite video: A video signal whose sync, color, and black-and-white signals are combined into one line. Examples are NTSC, PAL, and SECAM.

Composite sync: Combined vertical and horizontal sync pulses.

Control point: The reference point that controls the shape of a spline.

Convergence: The ability for a monitor display's red, green, and blue components to meet at the same location to create white.

Depth cuing: A method of representing depth by illuminating the "far" parts of objects with less brightness or contrast than the "near" parts.

Diffuse light: Light that is scattered in all directions.

Digital signal: A signal whose amplitude can be only one value (voltage) or another, that is, either on or off.

Digitizing (video): "Grabbing" a particular frame of video for translating into a digital representation for computer manipulation.

Display list: A method of holding information about a digital object for rapid calculating and display.

Dissolve: A transition in which an existing image is gradually replaced by another.

Dolly: A camera's movement toward or away from a scene.

Dot pitch: The distance between the small dots on a color monitor. A smaller dot pitch means a finer, more detailed image.

Dynamic range: The range of the smallest amplitude to the largest. A high-contrast photograph has a smaller dynamic range than a low-contrast image.

Ease-in/Ease-out: The gradual transition in the rate of change in an animation (usually speed).

EDL: Edit decision list. A list of videotape locations for editing particular transitions and cuts. Used for final video editing.

Environmental map: An image depicting the scene from an object's viewpoint, used for simulating the "environment" around the object.

Extruding: Giving a 2-D object (having *x* and *y* dimensions) a dimension of depth (*z*), making it into a 3-D object.

Faceted: The appearance of an object having many flat surfaces instead of a smooth or rounded look.

Field: In video, half of a frame (with NTSC, a field is 262.5 horizontal lines at 59.94 hertz).

Film recorder: An analog or digital device for transferring computer images onto film.

Flip book: A method of viewing animation by first saving the images in RAM and then viewing them sequentially in real time.

Floating-point math: Calculations that use numbers that have fractional parts, such as 1.00, instead of 1, which is an integer.

Forced perspective: A camera's view that is extremely wide angle.

Frame: A single image which is part of a sequence. In video, 2 fields constitute 1 frame.

Gamma correction: With monitors, a method for improving visual appearance by correcting nonlinearities in display of color.

Genlock: The ability for a graphics board to accept the sync signals from another source, such as a camera or VCR.

Generations: In video, multiple copies of the original video source.

Gouraud shading: A medium-speed shading technique in which the shading value is determined at the center of polygon.

Granularity: The appearance of dots or segments instead of smooth gradations of a shade or color.

Heliometric: The ability to orient the light (and shadows) of a 3-D model with the correct position of the sun based on latitude, time of day, month, and year.

Hidden lines: When displaying a 3-D wireframe model, lines behind the surface faces are not seen, as if the faces were opaque.

Hue: The shade of a particular color.

Hz: Cycles per second. Megahertz is million cycles per second.

Interlacing: A method whereby a video image is composed of alternating odd and even scan lines instead of sequentially displayed scan lines.

Interpolation: The calculation of values (or locations) in between two other values.

Keying: A process of selectively overlaying one video signal on another.

Keyframe: A particular frame in an animation that is created as a user-determined guide for interpolation to the next keyframe.

Loopthrough: A method for cabling one device that provides connectors to another device for sharing signals.

Luminance: The amplitude of the black-and-white portion of the TV signal.

Mapping: A method for placing an image on a surface of an object to create a realistic appearance.

Metamorphosis: Changing an object from one shape to an entirely new shape over time.

Moire: A pattern caused by the interference of the mixing of two separate patterns.

Motion blur: A method for mimicking an effect of rapidly moving objects by blurring their prior positions.

Normal perturbation: Disturbance of the dimensions of a face (the normal) of a 3-D object. Used to change the curved surface of a sphere into a deformed nugget.

NTSC (National Television Standards Commission) video: Also known as composite video and RS-170A. The signal standard that is used for television broadcasting.

Object: Surfaces that are grouped together for purposes of shading, moving, or building new objects.

Offline editing: Initial editing done on duplicate video tapes.

Online editing: Final editing session with the master tapes, usually executed via an edit decision list.

Opacity: The quality of a material to obstruct the viewing of objects behind it.

Orthographic projection: A method of viewing so that the distance of the viewer to the object doesn't change the appearance of the object.

Overscanning: With video displays, a term used to describe a video image that is scanned so that you see only the inside of the video image.

POV: Point of view, usually from the camera.

PAL: Phase Alternating Line. A European video standard that has 625 lines at 25 frames per second.

Pan: To rotate a camera's position around its vertical axis.

Patch: A mathematically defined curved surface.

Phong shading: A relatively slow method of shading in which calculated values are based on the angle of the face of a polygon.

Pin registration: A method for assuring absolute stability and repeatability of the position of film in the camera of a film recorder.

Polygon: An object consisting of a set of three or more straight lines.

Post-production facility: A company that provides services such as video editing or any of the services required after videotaping.

Preroll: The process of preparing the videotape to attain the proper speed when used in a VTR/VCR for editing or recording.

Primitive (graphics): A simple geometric object, such as a cube or sphere.

Radiocity: A method of rendering that creates accurate photorealistic images but consumes a great deal of computational time.

RAMcorders: A videotape recorder of sorts that uses RAM—electronic memory—instead of videotape.

Random access: In editing or computer backup, the process of getting to a particular location of videotape or hard disk via a user-chosen address. Its counterpart is serial access.

Ray tracing: A rendering method based on tracing the origin of light for each pixel of each object. Its rendering time is very slow yet produces photorealistic images.

Real time: The viewing of an event as it is happening.

Reflection map: A method of mapping an image onto an object in which the image is a rendering of the object's environment. The resulting mirror effect is used for added realism.

Render: To take 3-D model information and its associated values (camera position, lights, and so on) and create an image on the computer screen.

Resolution: The amount of detail that an image or display possesses as measured in x- and y-values.

Resolution independent: The ability to create images or 3-D objects that have no fixed resolution and that can be displayed at any arbitrary resolution after their creation.

RGB: The primary colors red, green, and blue.

RS-170A: The electrical specification for color video in NTSC.

Saturation: The amount of color in an image (having no black-and-white value).

Scan converter: An electronic device that typically converts high-resolution displays into recordable video signals.

Scan line: A horizontal line of a graphics display, usually video.

SECAM: Sequential colour à memoire, a French video standard.

Serial access: In editing or computer backup, a method of getting information serially, that is, examining all of the data sequentially until the proper location is obtained.

Single-frame animation: The process of creating animation one frame at a time, as opposed to seeing the animation run as it is intended to.

SMPTE: Society of Motion Picture and Television Engineers. A trade organization that initiates standards.

Specular highlights: Lighting highlights in which the reflection is concentrated at the source of light, making the object appear shiny.

Spline: A mathematically (polynomial) defined curve used instead of polygons for describing 2-D and 3-D objects.

STA: Safe titling area. The (80%) region within viewable video that will assuredly display the graphics information.

Surface of revolution: An object made by rotating a 2-D polygon around an axis, such as a vase made from half of its outline.

Switcher (video): An electronic device for routing video signals. Modern units, however, have much greater capability.

Sync (synchronization): The coordination of timing signals.

Tape weave: A defect seen with motion picture films where the film's position weaves or drifts about.

TBC: Time base corrector. Electronic device that corrects video timing errors.

Telecine: A system for transferring motion picture films to video.

Temporal aliasing: Such unwanted artifacts as wagon wheels appearing as though they are going in reverse when seen in feature films. It results from the process of capturing the image in discrete segments of time.

Termination: An electrical method for ending a chain of signals with resistance, usually via a switch.

Texture map: A 2-D digital image used for placing onto a 3-D object, giving it texture, thus creating a greater appearance of realism.

Throughput: The rate of data being transferred.

Tilt: To rotate a camera's position upward or downward on its horizontal axis.

Time code: A code, usually recorded on an audio track on a video tape, that defines the explicit location of the tape based on its elapsed hours, minutes, seconds, and frames.

Truck: To move a camera so that it follows the action sideways, as though it were in a truck keeping pace with the action.

Tweening: Creating in-between images based on a beginning and ending image.

Underscanning: A video term used with monitors to indicate that the video image includes the surrounding black border, thus displaying all of the available video image.

VITC: A time coding method in which the information about hours, minutes, seconds, and frames of the video tape is placed on the outside of the visible region of the video signal.

VGA: A graphics board and standard (of sorts) for IBM PC/ATs that supports a resolution of 640×400 with 256 colors of a 256,000 color palette.

Visual element: A component of the total visual image or animation.

WORM: Write Once/Read Many. A type of optical storage device used to store large amounts of data.

YIQ: The luminance and color difference signals in the NTSC video standard.

INDEX